Principles of VLSI RTL Design

Sanjay Churiwala · Sapan Garg

Principles of VLSI RTL Design

A Practical Guide

Foreword by Mike Gianfagna

Springer

Sanjay Churiwala
c/o Atrenta (I) Pvt. Ltd.
A-9, Sector 4
Noida – 201301
India
rchuriwala@hotmail.com

Sapan Garg
c/o Atrenta (I) Pvt. Ltd.
A-9, Sector 4
Noida – 201301
India
sapan@atrenta.com
sapan@sapanonline.com

ISBN 978-1-4899-9545-2 ISBN 978-1-4419-9296-3 (eBook)
DOI 10.1007/978-1-4419-9296-3
Springer New York Dordrecht Heidelberg London

Softcover re-print of the Hardcover 1st edition 2011

Printed on acid-free paper

Springer is part of Springer Science+Business Media (www.springer.com)

Foreword

It started slowly at first. The signs of change could be seen in the computer on our desk and then in our briefcase; in the car we drive and the gadget in our pocket that we used to simply call a cell phone. Electronics has undergone consumerization. Those complex things we used to call "computer chips" are now in everything. They make a connected world possible. They save lives and enlighten our children in ways we couldn't have dreamed of a short 10 years ago. The changes we've seen in our lifetime qualify as a revolution.

And the revolution has created some superstars along the way, in ways that are surprising when you consider the world of just a few decades ago. Engineers have become rock stars. The likes of Bill Gates and Steve Wozniak are the leaders of the revolution, and we all revere them. Geeks do indeed rule.

If we follow the revolution analogy a bit further, you will find that the armies of the revolution are composed of many thousands of design engineers, cranking out new ideas and new chips every day. They are the unsung heroes of the revolution. Not everyone gets to hold a press conference for a new product or throw your own party for several thousand of your closest friends, but the contribution of the world's design engineers is undeniable. Like any army, the soldiers of this one are specialized across multiple disciplines. Some focus on manufacturing, some on physical design and some on software and architectures.

While everyone has their part, and no chip can be made without contributions from all, there is one particular group of engineers that has a special place for me. It's the design engineers who sit between the next great idea and the implementation of that idea. These are the folks who translate the next big thing into a design description that can become the blueprint for the next product. They deal with highly abstract concepts on one hand, and deliver a robust plan to implement those concepts on the other. They are the register transfer level (RTL) designers, and they are people that Sanjay Churiwala and Sapan Garg have reached out to in this book.

Principles of VLSI RTL Design: A Practical Guide is destined to become a benchmark reference for this group of design engineers. The book treats a broad range of design topics in a way that is relevant and actionable for the RTL designer. As the book points out in many places, the decisions made by the RTL designer can have substantial impact on the overall success or failure of the chip project. Just as the hand of a talented architect can guide and influence a construction project, so can a talented RTL designer guide and influence a chip project – if they know how their decisions will impact the downstream tasks of implementation.

This books aims to educate and inform the RTL designer to understand just how powerful they can be. Thanks to their years of experience studying design and helping others to do it better through Atrenta's products, Sanjay and Sapan bring a wealth of relevant, actionable information to the table. I highly recommend this book for anyone who is part of the revolution, or aspires to be part of it.

San Jose, California, USA Mike Gianfagna

Preface

Dear Friends,

The idea and the inspiration behind writing this book did not come as one single "Eureka" moment. It got brewed within us for long. During many years of interaction with design managers and engineers at various IC design houses, we realized the importance and criticality of the role played by a good RTL designer in reducing the number of iterations from later stages to the RTL stage in the design cycle which helps the design to reach the market in time.

Specialized topics, such as DFT, Timing, Area, Power etc. have their own experts and are considered to come into play – much after RTL. However, the quality of RTL has a significant impact on these requirements. The domain experts of these specialized topics cannot be present at all times to guide the RTL designers. Many times, if an RTL designer is aware of what will cause trouble to these specialized stages later in the flow, he can at least consult with the specific domain experts, and, together they can judge on what would be best to do at the RTL stage itself. But, how can we make an RTL designer aware of these specialized topics? Imparting knowledge to RTL designer is the only way out. So, we hope this book will explain the fundamental concepts of all these specialized topics which an RTL designer should know – on the various impacts that his RTL has – on later stages of the design cycle. The book does not attempt to replace the domain experts. It tries to complement them – so that they can focus on the more complex things, while, explaining relatively simpler things is done by this book.

As part of our job at Atrenta, we have been receiving and studying RTL coding guidelines from many IC design houses. A lot of those rules needed us to analyze and think through as to what might be the main motivation behind the specific guideline. Sometimes, after the rules were coded into our software, and were in use at the design houses, we would get queries from the users as to why a specific guideline was important, or, what was the implication if they did not follow that guideline. Sometimes, it would be accompanied by: "We are aware of this RTL coding guideline that we are violating, and, we understand why this guideline is important. But, we have taken this alternative precaution. Do you still think, there is a trouble with our code?"

As we used to debate and discuss these queries within our company, and, many times with the users at the design houses, we started realizing that our users did not only expect us to just automate the rule checking but also they were looking at us – to actually define good coding practices. Though, most large design houses have a set of good design practices; such guidelines are missing at smaller design houses. And, even at places, where the designers have access to such coding guidelines, many times the reasoning and implications are understood by the more experienced designers, while, the relatively inexperienced designers are simply expected to follow them. These designers also want to understand the reasoning behind those specific guidelines.

An RTL designer should be able to appreciate that the guidelines are not to prevent him from expressing his creativity; rather the guidelines are to prevent his RTL design from running into trouble at a later stage. The above thought process was getting inspired by another important fact of ASIC design process. ASIC design is not about putting in a bunch of bright and smart engineers on a project and letting them do the design. Rather, ASIC design has a huge element of being able to foresee the downstream impact of their specific code. It is less to do with smartness and intelligence; and more to do with "knowing". Obviously, "knowing" would come with experience. And, it can also come with reading from other people's experience. This is what this book is trying to do.

Hopefully, this book will find its place in the hearts and minds of anybody who generates RTL code. This includes RTL designers as well as those writing tools that generate RTL. Relatively new RTL designers will find this book to be an interesting, rich and useful collection of knowledge at one place. Experienced RTL designers will be able to appreciate and cement some already known concepts. Domain experts can enjoy the reduction in routine queries and concentrate on more complex matters in this domain. We expect their continued guidance in terms of improving the material further – for future.

Acknowledgments

Though, the book lists two of us as authors, there are many people who have contributed majorly towards the creation of this book.

At the outset, we would like to thank, Dr. Ajoy Bose, Mike Gianfagna and Sushil Gupta of Atrenta for having provided us the support and encouragement to go ahead with this book. We would also like to thank Charles Glaser (of Springer) for providing us guidance at all stages.

We would also like to thank Mithilesh Jha (of Masamb), Ashutosh Verma (of Cadence), Ajay Singh (of Synopsys), Kevin Grotjohn, Wolfgang Roethig and Shabnam Zarrinkhameh. They influenced our understanding of many of these topics – during early stages of our career.

We would also like to thank Paresh Joshi (of Texas Instruments), Bhaskar J. Karmakar (of Texas Instruments), Jaya Singh (of Texas Instruments), Sandeep Mehrotra (of Synopsys), Ashok Kumar (of Oracle) and Gilles GUIRRIEC (of ATMEL) for having reviewed specific sections of the book.

A lot of the contents of the book has evolved based on lots of discussions with our colleagues (at Atrenta), and, many designers at our customers, and, so, we would like to thank them too. Charu Puri spent a lot of her time – in helping us with the graphics.

A special mention goes to Ken Mason (of Atrenta) – who during one breakfast casually mentioned to us, "Why don't you write a book – based on your knowledge on RTL design?" That was our first inspiration. It took several years after that.

And, at one place, when we were pretty much stuck, Prof. Manish Sharma (of IIT Delhi) gave us the push – which got us over the hump.

And, last but the most important one, we would remain indebted to our families. This being our first experience in authoring a book we had been spending much less time with our families. But, instead of complaining they kept motivating us. And, our kids – Arjun and Radhika Garg; Ruchi and Lubha Churiwala were equally excited at this venture of ours. Their follow up on our progress was more regular than the publishers☺ And, that did keep us going.

Happy Reading!

Noida, India
December 2010

Sanjay Churiwala
Sapan Garg

Contents

Chapter 1
Introduction to VLSI RTL Designs

For a moment, think of all the devices you use in your daily life e.g. mobile phone, computer, TV, car, washing machine, refrigerator, ovens and almost everything, even a smart card, credit card or a debit card. All these have one thing in common and that is, one or more semiconductor chips, also known as Integrated Circuits (IC). An IC comprises of millions of transistors etched on a small piece (say 10 mm x 10 mm) of doped silicon (a widely used semiconductor) with appropriate interconnections between them depending upon the desired functionality. Just compare this design complexity with the era of vacuum tubes where one vacuum tube was used to do the same functionality of a single transistor. Some of you may very well have seen the old age radio receiver box back where five or six such vacuum tubes could be seen lighting. Naturally, even that would be considered complex in those days relative to the technology and tools available in those times. All the design and manufacturing used to be mostly manual.

Today, the technology has advanced and automatic tools are available. In terms of technology, we can now fabricate millions of transistors in a unit square inch piece of silicon and therefore this technology is popularly known as Very Large Scale Integration (VLSI). To facilitate this complex design and fabrication of an IC, various automatic tools and machines are available. Design and verification of an IC is done on very fast computers with the aid of Electronic Design Automation (EDA) software tools.

1.1 A Brief Background

Typically, an IC has two major sections.

- *Analog* – this section of an IC generally interacts with the real world and uses all voltage levels of a signal. For example, section receiving an RF signal, peripheral PAD circuitry, PLL section, etc.
- *Digital* – this section forms the core of an IC and mainly deals with two levels (1 and 0) of a signal. All data transfer between various sections, data processing and all computations are accomplished using digital design. For

S. Churiwala, S. Garg, *Principles of VLSI RTL Design*,
DOI 10.1007/978-1-4419-9296-3_1, © Springer Science+Business Media, LLC 2011

example, all the datapath (ALU, data transfer between various sections, etc), memories (RAM and ROM), control circuitry (clocks, enables and set/reset handling) etc. This book deals with the digital section of an IC and unless otherwise specifically mentioned rest of the book implicitly talks only about digital design.

EDA tools started emerging in late seventies. Mainly, it started with using some mechanical design software for layout of the electronic designs on a Printed Circuit Board (PCB). However, the major breakthrough was achieved in mid eighties with the advent of Hardware Description Languages (HDL). Verilog was the first one to get off the block and immediately thereafter came VHDL. Having the circuits represented in terms of HDL meant that there could be automation done in various aspects of circuit design e.g. functional verification of the circuit, mapping the HDL into logic gates, etc.

1.2 Hardware Description Languages (HDL)

There are two most popular and widely used hardware description languages in semiconductor and EDA industry:

- *Verilog* – first version came in 1985 and since then there have been several other versions like 1995, 2001, 2005 and SystemVerilog.
- *VHDL* – first version came in 1987 and since then there have been several other versions like 1993, 2000, 2002, 2006 and 2008.

Using HDL, there are three levels of abstraction possible to define:

- *Behavioral Design* – in terms of algorithms of the design. This is the highest level of abstraction possible using the HDL. Here the top level function of the design is described and it is useful mainly at the system analysis, simulation and partition stage. Contrary to RTL design, behavioral design may or may not be synthesized into logic gates automatically (by synthesis tool).
- *RTL Design* – in terms of data flow between registers (storage elements) of a circuit. This is a much higher level of abstraction than netlist description of a circuit but still has detailed description of the circuit with respect to the clocks and data flow. So, RTL design comes in between behavioral and netlist (explained in next point) as far as the abstraction of a design is concerned. This is the most widely used form of any HDL by a hardware design professional. The following example shows a sample Verilog RTL description:

```
module dff_rtl (data, clk, reset_n, q);
input data, clk, reset_n;
output q;
```

```
reg q;

always @ (posedge clk or negedge reset_n)
if (~reset_n) begin
  q <= 1'b0;
end else begin
  q <= data;
end

endmodule
```

Here a D flip flop has been designed using procedural construct *always* and conditional construct *if-else*. Subsequently, synthesis tools are used to convert the RTL description into a netlist as shown in Fig. 1.1.

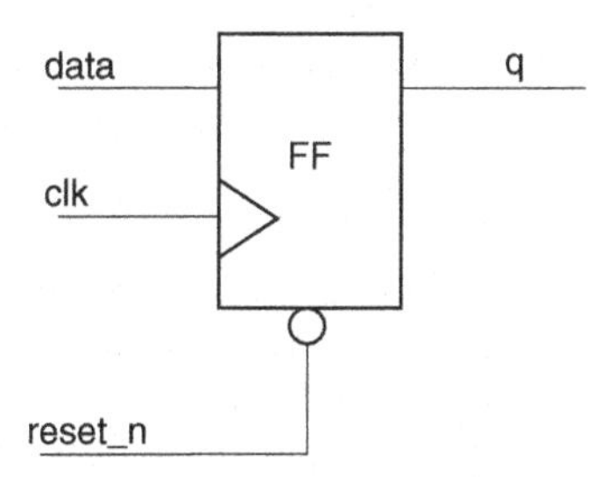

Fig. 1.1 RTL code synthesized into a D flip flop

This book is on RTL designs only and hence all aspects of RTL will be discussed in details in subsequent sections and chapters.

- *Netlist Design or Gate level Design* – in terms of instantiations of cells from a library. This is closer to physical representation of a circuit and has almost zero abstraction. This form of HDL is generally the output of a synthesis tool which takes RTL design as input. The following example shows a Verilog netlist design:

```
module half_adder_netlist (x, y, sum, carry);
input x, y;
output sum, carry;

and inst_carry (carry, x, y);
xor inst_sum (sum, x, y);

endmodule
```

As evident from the above description and Fig. 1.2, one instance of *and* and one of *xor* primitive is used to design the half adder.

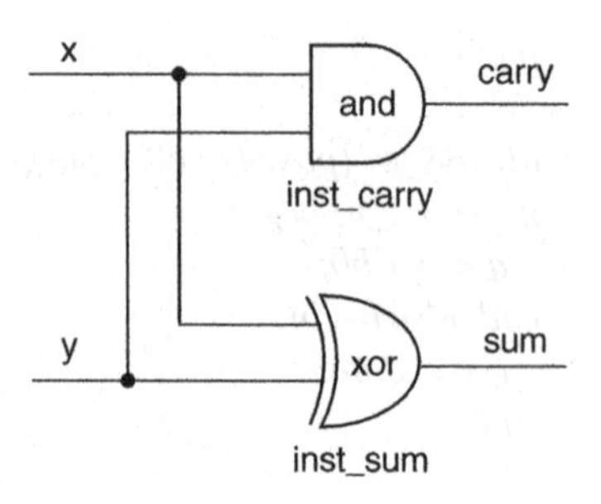

Fig. 1.2 Synthesized version of a gate level design

1.3 RTL Designs

Register Transfer Level (RTL) description is a lower level of abstraction than behavioral description but higher level abstraction than netlist description of a circuit design. As shown in last section, RTL description is written using either Verilog or VHDL (using any of their flavors). In RTL description, circuit is described in terms of *registers* (flip-flops or latches) and the data is *transferred* between them using logical operations (combinational logic, if needed) and that is why the nomenclature: Register Transfer Level (RTL).

RTL description is the first real dig at the detailed circuit design. All the subsequent steps in IC design process depend upon the quality of the RTL design. Good quality RTL means its easily realizable, correct, reusable and suitable for downstream tools in the flow. Better the quality of the RTL design, the quicker it would be to get the IC into the market. "What you do up front determines how difficult and costly your development flow will be" said Gabe Moretti in his article "Increasing the level of Abstraction of IC design" on EDACafe. So, it is amply clear to one and all in semiconductor industry that RTL design quality is very important and this is where this book might prove useful.

Consider the following example of a simple RTL design and its quality.

```
module top (in, clk1, clk2, out);
input in, clk1, clk2;
output out;
wire q1,q2;

flop f1 (in, clk1, q1);
flop f2 (q1, clk2, q2);
flop f3 (q2, clk2, out);

endmodule

module flop (d, clk, q);
input d, clk;
output q;
```

```
reg q;

always @ (posedge clk)
q <= d;

endmodule
```

For the above RTL code, the synthesized gate netlist would be something like Fig. 1.3.

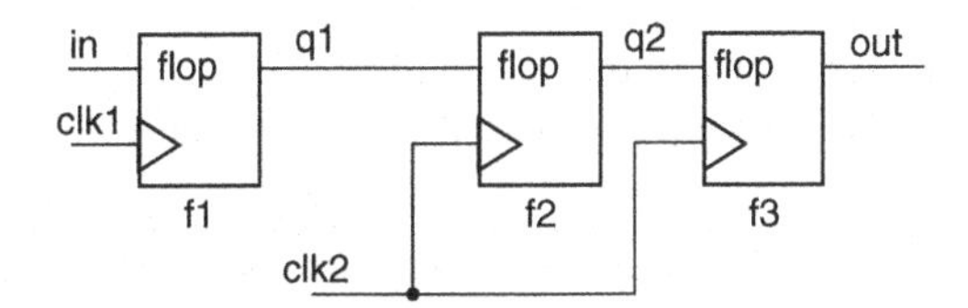

Fig. 1.3 A synthesized version of RTL with clock domain crossing

Here you can see that data from input port *in* is captured in flop *f1* at positive edge of clock *clk1*. And then it gets captured in flop *f2* at positive edge of clock *clk2* and then finally reaches the output port *out* through flop *f3* at positive edge of clock *clk2*. So, the data crosses the clock domain from *clk1* to *clk2*. Note the double flops used in clock domain *clk2* before transferring the data finally to output port *out*. What was the need of double flops when apparently a single flop was enough? Since *clk1* and *clk2* are asynchronous to each other, the extra flop gives sufficient time for signal coming from *clk1* domain to stabilize in *clk2* domain before it is finally presented at the output. This is a good quality RTL as appropriate synchronizer (here double-flop synchronizer was used) has been used at a point where signal crosses clock domains. The example here was just for illustration of good quality RTL and everything about clock domain crossings, their characteristics, various synchronizing techniques, etc is presented in Chapter 4.

1.4 Design Goals and Constraints

All electronic systems need chips inside it. And since there are many system companies competing to win the market share for a specific kind of system, they have to build their system as early as possible and everyone tries to reach the market before others. So, the system companies give the chip requirements and an aggressive schedule to chip companies. Naturally, there is competition amongst chip companies too. So, a system company would choose a chip company which can deliver a quality chip in most aggressive schedule. Hence, the most important goal of any chip design is to deliver a *quality design* in shortest time. To achieve this, the entire project is broken into several sub-goals or milestones like architecture completion, high level chip design, front-end completion and back-end completion.

Now what is a *quality design*? A quality design is a design which works meticulously well with the given *design objectives and constraints*. Following are the major design objectives and constraints:

- Functionality of the design (design objective)
- Speed at which a design can operate (design constraint)
- Power consumed by the design (design constraint)
- Testability of the design (design objective)
- Area of chip package (design constraint)

These top level chip constraints are further broken down into many block level constraints. Several trade-offs have to be made to ultimately make the design work – respecting all the top level constraints. So, these constraints present many challenges to the designers at each stage of the design flow. If an RTL designer is aware of these challenges and tunes his RTL to help lower stages to meet these constraints then the likelihood of completing a quality design in time increases. Following sections and chapters of this book will highlight these challenges and will try to educate an RTL designer on all these concepts.

1.5 RTL Based Chip Design Flow

IC design is a complicated process involving several steps of equal importance. To remain competitive and successful in market, it is of utmost importance for all semiconductor design companies to have a well documented and tested design methodology for this complicated process to deliver ICs in time. A simplified version of a typical IC design methodology is shown in the following flowchart.

As evident from the design methodology, most of the design steps are performed after the RTL has been written and analyzed. Any quality issue in RTL would cause issues in subsequent steps resulting in costly iteration back to RTL design step. Each such iteration consumes a lot of precious time. And there may be several such iterations because of several uncaught issues in RTL design. In fact, some of these issues may well go unnoticed altogether resulting in faulty chip in a real system in the hands of a user. There are examples of famous semiconductor companies recalling the faulty chips from the field and replacing it for free. Understandably, apart from embarrassment, a lot of money is also spent in this damage control and recovery process.

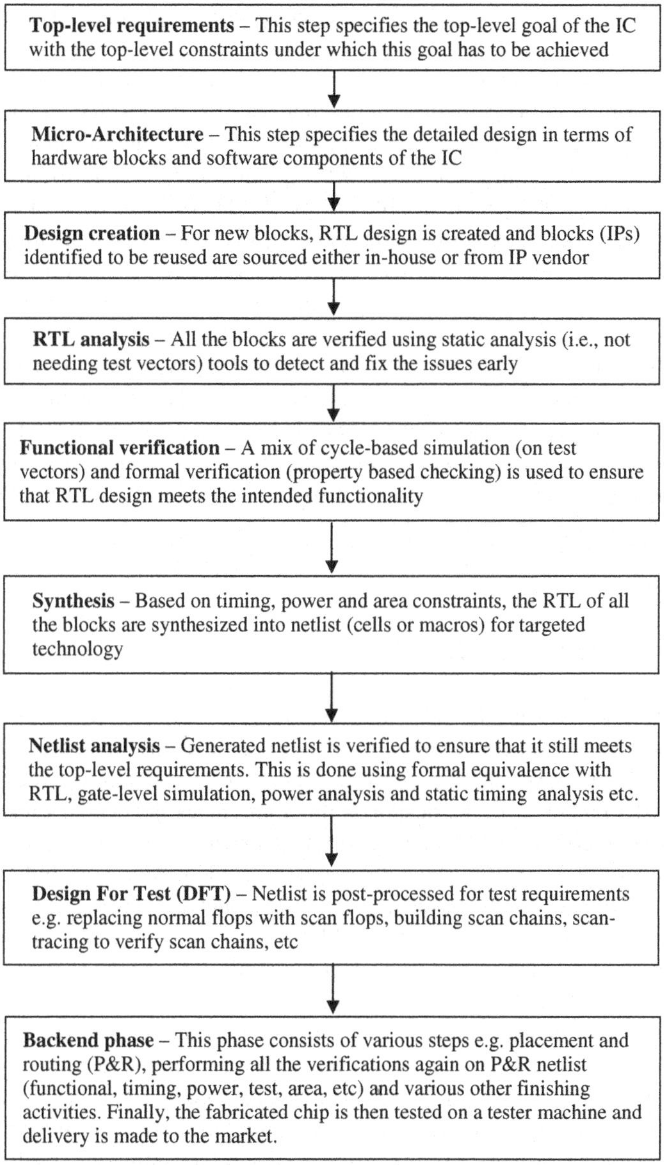

Top-level requirements – This step specifies the top-level goal of the IC with the top-level constraints under which this goal has to be achieved

↓

Micro-Architecture – This step specifies the detailed design in terms of hardware blocks and software components of the IC

↓

Design creation – For new blocks, RTL design is created and blocks (IPs) identified to be reused are sourced either in-house or from IP vendor

↓

RTL analysis – All the blocks are verified using static analysis (i.e., not needing test vectors) tools to detect and fix the issues early

↓

Functional verification – A mix of cycle-based simulation (on test vectors) and formal verification (property based checking) is used to ensure that RTL design meets the intended functionality

↓

Synthesis – Based on timing, power and area constraints, the RTL of all the blocks are synthesized into netlist (cells or macros) for targeted technology

↓

Netlist analysis – Generated netlist is verified to ensure that it still meets the top-level requirements. This is done using formal equivalence with RTL, gate-level simulation, power analysis and static timing analysis etc.

↓

Design For Test (DFT) – Netlist is post-processed for test requirements e.g. replacing normal flops with scan flops, building scan chains, scan-tracing to verify scan chains, etc

↓

Backend phase – This phase consists of various steps e.g. placement and routing (P&R), performing all the verifications again on P&R netlist (functional, timing, power, test, area, etc) and various other finishing activities. Finally, the fabricated chip is then tested on a tester machine and delivery is made to the market.

1.6 Design Challenges

It is evident from the previous sections that the quality of the RTL directly impacts the success and timeliness of the subsequent steps, including the ability to achieve the desired function, electrical characteristics, manufacturability, fault detection etc.

of the IC. The RTL lays down the foundation of an IC and hence presents many challenges for an RTL designer.

As an analogy, consider an engineering drawing of a house that an architect creates. If something is wrong with the engineering drawing, the result could be that the house is not structurally sound, and thus might require dismantling and re-building. RTL design is similar. An error in the RTL might not be visible to an engineer immediately, but it could create trouble during the downstream steps of IC design methodology, resulting in schedule delays and a lot of rework. If one either has the experience and knowledge to spot these problems early, or software tools to help analyze the RTL to make sure that it is error free, then the entire design process will go much more smoothly with minimum iterations back to RTL.

The goal of this book is to provide the practical knowledge – so that an RTL designer can understand the downstream impact of the RTL. It explains the various aspects, their significance, and what care needs to be taken during RTL design and why. This section will provide a quick look on various design challenges here without really getting into a maze as there are dedicated chapters on each of these giving all the details and fundamentals.

1.6.1 Simulation Friendly RTL

Immediately after writing the RTL and analyzing it statically, it has to be made sure that the design's functional intent is met. To achieve this, an appropriate HDL simulator tool is used which takes test cases (popularly known as test vectors in IC design industry) along with the design as input and gives the design's response as output. Designer compares this output with the expected results and reiterates with the RTL in case of any difference between the two. Generally, with little iteration between the RTL and simulator, the design intent is met and designer gets satisfied. These iterations between the simulator and RTL are not that costly as this is just the second step after the RTL writing which means only RTL analysis step has to be repeated between RTL change and simulation re-run.

Achieving functionally correct RTL is only the first small step forward. Next in line is to ensure that even the gate level design created (from the functionally correct RTL) by appropriate synthesis tool also matches the intent. To do this comparison, same test vectors are given to the gate level simulator along with the synthesized gate level design. In case there is any difference found in simulation results of RTL design and corresponding synthesized gate level design (popularly known as *pre and post synthesis simulation mismatch* in IC design industry), reiteration back to RTL is done to fine tune the RTL so that design intent is met equally at both RTL and gate level. These iterations between the synthesis output and RTL are costlier as: one, synthesis is a time consuming step and two, with each extra iteration RTL simulation also has to be repeated. So, it is better to avoid these iterations.

What is the reason of this pre and post synthesis simulation mismatch? It happens because for some specific pieces of RTL code the simulator interprets it in some way and the synthesis tool in a different way. Have a look at one such example piece of RTL code.

```
module fn_return_vals (in1, in2, sel, sum);
input [3:0] in1, in2;
input sel;
output [4:0] sum;

reg [4:0] sum;

always @ (in1 or in2 or sel)
  sum = Adder(in1, in2, sel);

function [4:0] Adder;
input [3:0] first, second;
input add_nibble;
case(add_nibble)
  1'b1: Adder = first + second;
// 1'b0: Adder = 5'b00000;
endcase

endfunction

endmodule
```

In the above RTL code, the function is not returning value when *add_nibble* is 0 because that part has been commented in the code. In such a case, simulator would assign 0 to the *Adder* whereas some synthesizers may assign an arbitrary value. So, there would be a pre and post synthesis simulation mismatch in results. To correct this, the designer has to reiterate back to the RTL and uncomment the commented portion because the intent of design was to have 0 and not an unknown value. Clearly, this costly iteration could have been easily avoided provided the knowledge that "for all possible states, a function should have well defined return value" is there with the RTL designer. Not only these iterations but any iteration from any subsequent step after synthesis is even costlier as more steps would have to be repeated for closure. Hence, it is highly desirable to have a good quality RTL upfront.

There are many such ambiguities possible in a RTL code and all of those have been explained with fundamentals in Chapter 2.

1.6.2 Timing-Analysis Friendly RTL

As you know, an RTL is synthesized into a netlist comprising of millions of logical devices/gates. Then, these gates are placed and interconnected with metal wires for a specific chip size. Naturally, the devices will have some intrinsic delays and metal wire connections will have some interconnect delays proportional to the length of the wire. So, there are two major components of delay which a signal encounters during its propagation inside the circuit:

- Device/Gate delay is the time it takes for a signal to change at output with a corresponding change in input.
- Interconnect delay is the time it takes for a signal to travel from an output of one device to the input of other connected device.
  In Fig. 1.4, *t1* is device delay and *t2* is interconnect delay.

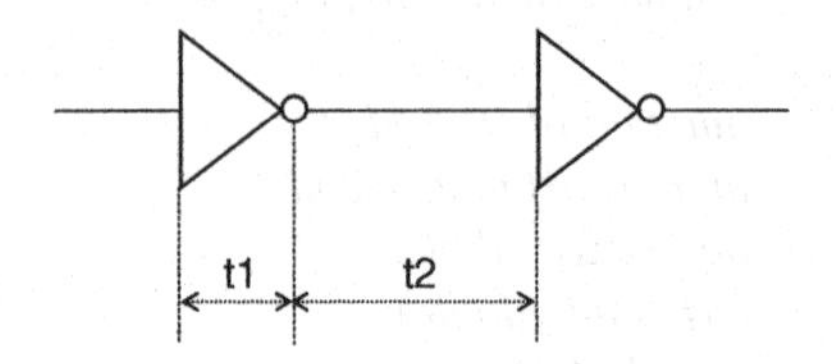

Fig. 1.4 Device delay (*t1*) and inter-connect delay (*t2*)

At RTL stage, it is very difficult to know the actual delays because:

- gate delays can only be known once the RTL has been synthesized and mapped to a specific library
- and, interconnect delays can only be known once the devices/gates of the synthesized netlist have been placed and connections between them have been routed
  Hence any kind of accurate timing analysis on a design can only be done after synthesis, placement and routing.

Why this timing analysis is needed? Most of the operations inside an Application Specific IC (ASIC) are synchronous to a reference signal, popularly known as *clock* signal. The frequency of *clock* signal defines the speed at which an IC operates. Depending upon the application, an IC is designed for a certain clock frequency. For example, a typical microprocessor of a PC has a speed of 2–3 GHz. And, as you have seen an IC comprises of millions of gates and metal wire interconnections, a propagating signal comes across gate delays and interconnect delays. Because of these delays, it has to be ensured that a signal reaches specific points in the circuit within desired time otherwise speed of the circuit will have to be compromised for correct functionality. Hence, timing analysis of a design is mandatory.

But as explained above, any kind of accurate timing analysis can only be done after the design is synthesized. So, what can you do as an RTL designer to make your RTL timing friendly even before it is synthesized to avoid costly synthesis runs?

Your endeavor should be to write an RTL which not only meets the functional requirements but also is *timing-closure friendly*. An RTL can be timing-closure friendly in two ways. First is to follow some simple guidelines within the RTL which are sure to help in faster timing closure. For example, avoiding deep combinational logic for signals of interest, avoiding large shift registers, avoiding large fanout for signals of interest, etc. Second is to write the RTL in such a way that downstream timing analysis tools do not have to take any ambiguous/risky decisions to achieve timing closure. Look at one such specific piece of RTL code.

```
module avoid_comb_loop (in1, in2, out);
input in1, in2;
output out;
reg out, out1;

always @ (in1 or in2)
if (in1 == 1'b0)  out = in2;

endmodule
```

For the above RTL code, as shown in Fig. 1.5, a latch is inferred because there is no *else* defined for the corresponding *if* in the combinational block.

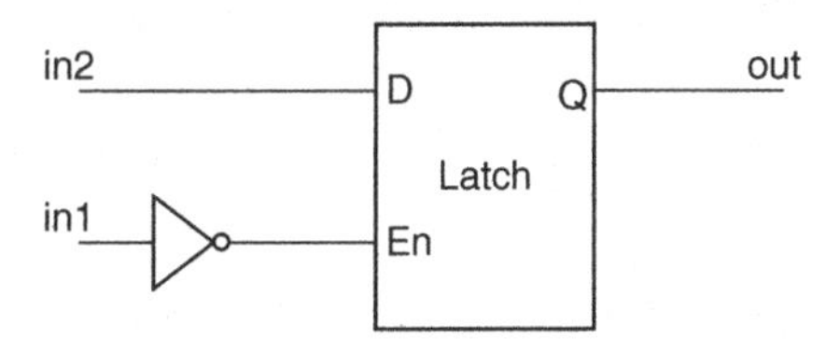

Fig. 1.5 RTL synthesized into a latch

Generally, it's a good practice to avoid latches especially the unintentional ones. This is because latches are more susceptible to noise as they are level sensitive and have intrinsic combinational feedback path in them. Following is the changed RTL code which will avoid the latch but now it leads to combinational loop in the circuit as shown in Fig. 1.6.

```
module avoid_comb_loop (in1, in2, out);
input in1,in2;
output out;
reg out, out1;

always @ (in1 or in2)
if (in1 == 1'b0)  out = in2;
else out = out;

endmodule
```

Combinational loops create problems for several downstream tools including Static Timing Analysis (STA). To do timing analysis, an STA tool has to break the

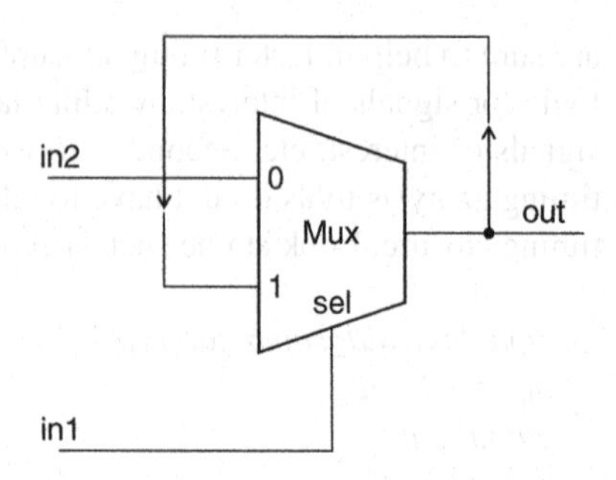

Fig. 1.6 RTL synthesized into an equivalent circuit with a combinational loop around mux

combinational loop somehow. Without any inputs from the designer, the STA tool will break it by deactivating one of the segments of the loop. For this deactivated segment, there will not be any timing analysis done by the STA tool. This could be a problem because this deactivated segment may have been a part of one of the critical paths of the design which will go un-analyzed now. Combinational loop can be avoided by changing the RTL code as shown below. The corresponding synthesized version is shown in Fig. 1.7.

```
module avoid_comb_loop (in1, in2, clk, out);
input in1,in2, clk;
output out;
reg out, out1;

always @ (posedge clk)
out1 = out;

always @ (in1 or in2)
if (in1 == 1'b0) out = in2;
else out = out1;

endmodule
```

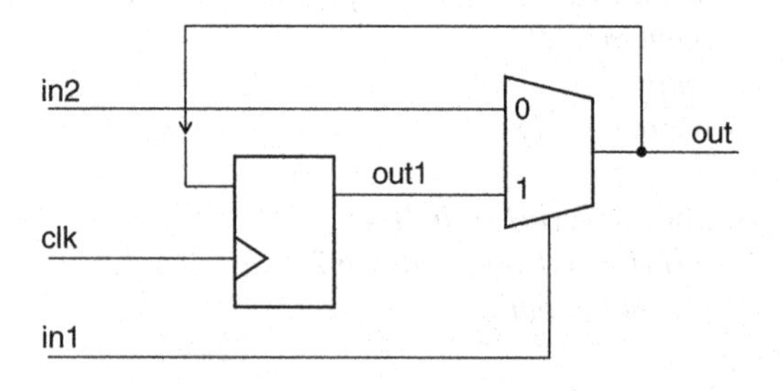

Fig. 1.7 RTL synthesized into an equivalent circuit without any combination loop or latch

There are many such precautions to be taken in a RTL code to make it more suitable for STA and timing closure and all of those have been explained with fundamentals in Chapter 3.

1.6.3 Clock-Domain-Crossing (CDC) Friendly RTL

Most of the ICs today are almost like entire *System on Chip* (SoC). This means there are several independent applications on the same chip. These applications may have different requirements with respect to the speed they have to operate on. As you will see later, higher the speed of operation higher will be the power consumed. So, applications which do not require high performance are operated at lower frequency and applications which require high performance are operated at higher frequency. Hence, there may be several clocks running at different frequencies on the same chip. So the challenge is whether your chip keeps all the data intact and maintains its integrity as it travels different applications running asynchronously and most probably on different clock domains. When data crosses from one clock domain to the other, it is called as *Clock Domain Crossing* (CDC).

In a CDC, the clock domain from which the data is generated is called as source clock domain, and the domain where it is captured is called as destination clock domain. For correct functionality, data generated in source domain has to be captured properly in destination domain, without it getting lost or corrupted in between because of CDC. Any CDC has potential to give rise to several data integrity issues, but the good thing is that there are known ways to mitigate these data integrity risks. So, what can you do to ensure that all the CDCs in your RTL have been taken care of well? Well, a lot can be done. Your endeavor should be to make the RTL not only simulation and timing-analysis friendly but also make it CDC friendly. Look at one of the most important issue which arises because of CDC and how you can take care of it in your code.

Metastability is the most common issue which could arise due to CDC. Simply put, a metastable state is a state where a signal is neither at level *0* or level *1* i.e. in an undefined state. Due to setup/hold timing violations at the first flop in destination domain, this flop could go into a metastable state, thus, causing chances of failure. With increasing frequencies at which the ICs operate these days, the chances of failures are more. So, metastability has to be handled very well for each flop in the design. In case of asynchronous CDCs, the metastability is handled by adding proper synchronizer at the point of domain crossings as shown in Fig. 1.8.

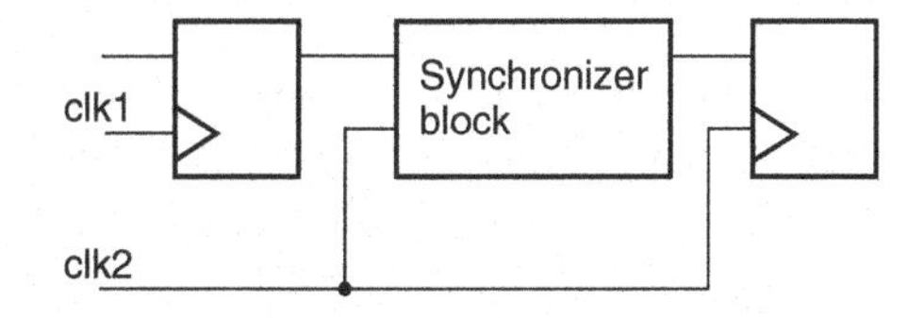

Fig. 1.8 Synchronizer to avoid metastability at CDC point

One such simple synchronizer is double-flop synchronizer which is generally used for CDCs of control signals in the designs. The example RTL code and corresponding figure presented in Section 1.3 is actually an RTL code which shows a CDC and a double-flop synchronizer used at the point of domain crossing.

Double-flop synchronizer is just one type of synchronizer but there are several other types of synchronizers at the disposal of RTL designer. Some of them are part of the library and you have to just instantiate the right synchronizer cell at the right point in the RTL. Some of these are double-flop synchronizer, synchronized common enable synchronizer, handshake synchronizer, fifo synchronizer, etc. Specific type of synchronizer is more suited for specific types of signals involved in CDC e.g. for control signals involved in CDC, a double-flop synchronizer is more suited whereas for data buses involved in CDC, a fifo synchronizer is more suitable.

There are bound to be CDCs in any design which you do today, so you need to make sure that the CDCs are handled well. Metastability, as mentioned above, is just one issue and there are other issues which can simultaneously exist on any CDC; e.g. data hold issue in case of fast clock to slow clock crossing and separately synchronized signals re-converging. All of these issues with remedies at the RTL stage itself are explained in details with all fundamentals in Chapter 4.

1.6.4 Power Friendly RTL

In today's world of wireless and mobile systems, it is very important to keep the power dissipation at the minimum possible. This helps in two ways. First, it extends battery life. Second, it will make the equipment less bulky by avoiding the need for exhaust fans for cooling.

An RTL is synthesized and mapped in terms of gates/devices which are in turn made up of basic building blocks called CMOS transistors. All ICs consume power because these transistors consume power in following conditions:

- When the load capacitors get charged and discharged (switching power). This is the major contributor of the total power consumed by an IC
- When the transistors are OFF (leakage power) as there is still some amount of current which flows through it because of reverse bias. At lower transistor geometries, the magnitude of leakage power is significant and is not far away from magnitude of switching power
- When both the transistors are ON momentarily during transition causing short-circuit between V_{DD} and GND (short circuit power)

So, how can an RTL designer help to keep the power minimum when power consumption is mainly a property of transistors of which an RTL designer does not have much visibility? Actually, a proper planning at even one level higher than RTL i.e. system design level, is much more desired. For example, portions of system which do not need high performance can be operated at a lower voltage or lower frequency levels. Even some portions of the system may be shut down when they are not in active use. Nevertheless, there are still ways in which an RTL designer can contribute.

The endeavor should be to write an RTL which is not only simulation, synthesis and timing friendly but also *power friendly.* An RTL can be power friendly in two ways:

- By ensuring that all power planning has been managed well i.e. portions of system operation at different voltages have been interfaced well through appropriate use of isolation cells and level shifters. Information about level shifters, isolation cells and other power strategies are given in a side file in UPF or CPF format. Synthesis tool will read this side file along with the RTL to properly place these strategically important power management related cells. So, an RTL designer has to ensure that this side file fully compliments the un-instrumented RTL. Meaning of un-instrumented RTL is that the RTL by itself does not have any power management cells' description.
- Most optimum logic is used to minimize the switching activity. For example,
  - Stopping the clock to reach the clock input of the flop for the duration in which either the D input of flop is not being changed (Fig. 1.9) or the output of the flop is not going to be used
  - To use left shifter for multiplication and right shifter for division instead of xor gates based multipliers/dividers
  - And many more techniques

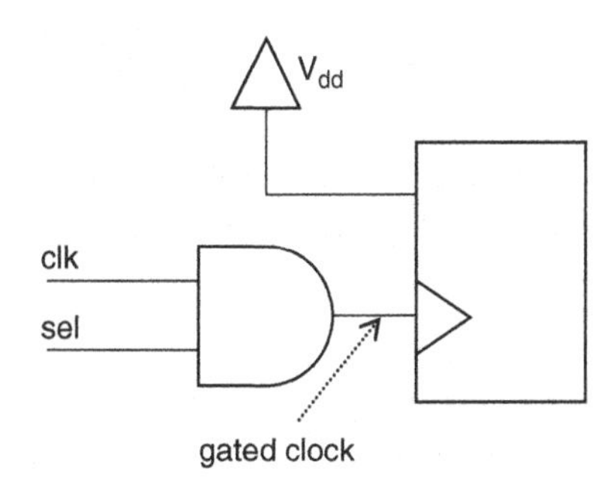

Fig. 1.9 Simplified version of clock gating to save power

Have a look at following piece of RTL code:

```
module gatedff_rtl (sel, clk, q);
input sel, clk;
output q;

reg q;
wire gated_clk;

assign gated_clk = clk & sel;
always @ (posedge gated_clk)
  q <= 1'b1;
endmodule
```

As you can see, input of the flop is tied to a constant. Naturally, output will remain at this constant value at all the time irrespective of the clock applied at its input. So, depending upon the design requirements user may either remove this flop altogether or if a flop is needed then gate its clock input so that clock does not reach flop's clock input. In this example, user has decided to gate the clock input. Please note that this is very simplistic hypothetical depiction of how clock is gated. In real situation, the flop's input will remain constant for a specific reason and period only and there will be some other considerations taken into account to properly gate the clock. But, to understand the concept this is sufficient and is depicted in the RTL code shown and corresponding Fig. 1.9.

Since, the clock input of the flop will not be switching now, it will save switching power. One may argue that now the input of AND gate to which clock is applied is switching instead of flop's clock input, so how it would make a difference as far as switching power is concerned. But the output of this AND gate might be driving several such flops. So, saving is at multiple places.

To ensure that you have written the RTL well to make it power friendly, you need a power estimation tool which can estimate the power at RTL. It is well understood that the estimated power at RTL may not be accurate as compared to power estimated after layout because parasitic (resistance and capacitance) of routed wires play a major role in calculating switching power. But, still some level of estimation at RTL which is fairly close enough to the post-layout estimation is very helpful for RTL designer to give him the confidence that he is designing his RTL which is power optimized.

All the above has been discussed and explained well in detail with all other power related fundamentals in Chapter 5.

1.6.5 DFT Friendly RTL

In today's competitive world, it proves almost fatal for a chip company if the chip delivered does not work in field. Therefore, there is absolutely no excuse for not testing each and every chip before it is delivered. The chips are tested on a very expensive *tester* machine. And, since this tester machine's time is very precious, the tests have to be done in an optimum way such that they unravel any possible defect in a minimum number of test vectors. Broadly, there are two kinds of tests: Functional tests and Manufacturability tests. Functional tests are done very extensively at design stage itself using simulators and various other analysis tools. Manufacturability tests, though, can only be done after the design is fabricated on a die. Manufacturability tests unravel the defects which creep in due to some impurities in the die or due to some issue in the fabrication process. Each and every chip manufactured in a foundry and delivered to a systems company, must be fully tested for any kind of manufacturing defects.

The manufacturing defects result in some nets either getting *shorted* with some other net or getting broken causing an *open*. So, test patterns are generated to catch these shorts and opens. Finally these test patterns are applied to the manufactured chip on a tester machine and if the results do not match the expected output, the chip is marked faulty and discarded. These test patterns are generated using some *fault modeling* techniques. One of the most commonly used techniques is *stuck-at fault* modeling. Two stuck-at faults are *stuck-at 0* and *stuck-at 1*. Each net in the circuit is tested whether it is stuck-at 0 or not. This is done by applying suitable values to its relevant input nets (if the net being tested is an output of a gate) and the effect is studied by looking at the nearest gate attached to it.

Similarly, each net is tested whether it is stuck-at 1 or not. As you can imagine now, for a multi-million gates design, the number of nets and thereby number of test patterns to be generated would be huge. So, the patterns for such testing are generated by Automatic Test Pattern Generation (ATPG) tools. As described till now, finding out manufacturing defects is entirely a late post-manufacturing step, so what can an RTL designer do for detecting these manufacturability defects? Well, an RTL designer can help a lot by writing an RTL which is friendly for not only ATPG tools but also other tools like scan insertion tools, scan stitching tools, etc. Don't worry about details of scan insertion/stitching etc for now as these will be explained in detail later in this book.

Let us see how an RTL designer can help an ATPG tool. The test patterns generated by ATPG tools are applied at the input pins of the chip and the results are observed at the output pins of the chip. Therefore, ATPG tools can only generate test patterns for those nets which are accessible from the input pins or scan flops (controllability) and whose values can be observed at the output pins or scan flops of the chip (observability). So, any net of the design will have all needed test patterns if it has clear controllability and observability. And, if a net does not have clear controllability and/or observability, ATPG cannot generate its test patterns. Hence, an RTL designer has to try and avoid some of the common issues in RTL which can hamper controllability and/or observability.

One of various such issues in RTL, which hamper the working of DFT related tools later in the flow, is having combinational loop in the design. Combinational loop has high probability of making some of the nets, in its fan-in cone, unobservable by blocking their path. So, an RTL designer is best advised to avoid combinational loop in his RTL. Similarly, having latches in the flop based design is not considered DFT friendly as special effort has to be put in making them transparent in testmode. Refer to the RTL examples shown in Section 1.6.2 where a combinational loop and latch is avoided. There are various other precautions which RTL designer should take to make his RTL DFT friendly like ensuring that asynchronous controls (set and clear of flops) are easily controllable in testmode, bypassing clock-gating enable in testmode, etc.

All the above has been discussed and explained well in detail with all other DFT related fundamentals in Chapter 6.

1.6.6 Timing-Exceptions Friendly RTL

An IC has to function well at a speed at which it is supposed to work. For example, a 2.7 GHz microprocessor should perform all its functions correctly at a clock speed of 2.7 GHz. To ensure this performance, the design has to be analyzed at all stages to meet timing. By default, data should take one clock cycle to travel from one flop to the next one. And, all data paths have to be analyzed to meet this constraint. This means a lot of work for timing analysis and physical synthesis tools as there are millions of such data paths in a chip. However, this work can be reduced if these tools are told about some paths for which meeting one cycle requirement is not applicable. These are called *timing exceptions*.

Specifying accurate timing exception is a very important task which needs to be done very carefully. It is a very important task because it gives direction to STA and back-end tools to not spend time unnecessarily on paths which are not critical to meet default timing of one clock cycle between the flops. And, this task needs to be done very carefully because even if one timing critical path is specified as timing exception path then the chip might fail in the field. Best stage to specify accurate timing exceptions is RTL stage. An RTL designer exactly knows his design's ins and outs so he is the best person to know which paths are timing critical and which are not. But, it is very important for him to specify timing exceptions very carefully. And that he can do well if he has good knowledge of all the concepts on what all can be treated as timing exception paths.

Timing exceptions themselves are not specified as part of RTL code. They are specified in a side file called as SDC file. Mainly, there are two types of paths specified for timing exception: *False Paths* and *MultiCycle Paths*.

False path is that path between two points of the circuit which is structurally present in the design but there is no requirement to meet specific timing (and hence would not hamper the normal speed of the system) for data to travel between these 2 points. An example of a false path between 2 flops is a path at asynchronous CDC as shown in Fig. 1.10.

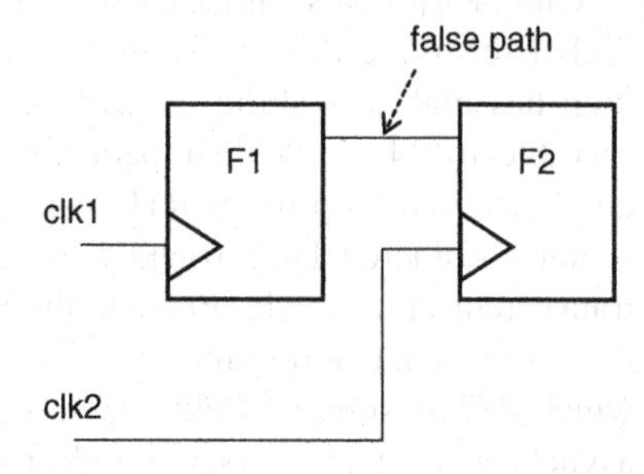

Fig. 1.10 False path at asynchronous CDC point

Since *clk1* and *clk2* are asynchronous to each other, there is no requirement for data launched by F1 to meet setup and hold requirements at input of F2. So, at asynchronous CDC point, RTL designer can specify the timing exception in SDC file.

MultiCycle paths are of two types. First type of multicycle path is that path between two points of the circuit which can take more than one clock cycle without hampering the intended speed of the system. One simple example of such a multicycle path is a path between two points when the source circuit has to wait for a data request signal from destination circuit before sending the next data as shown in Fig. 1.11.

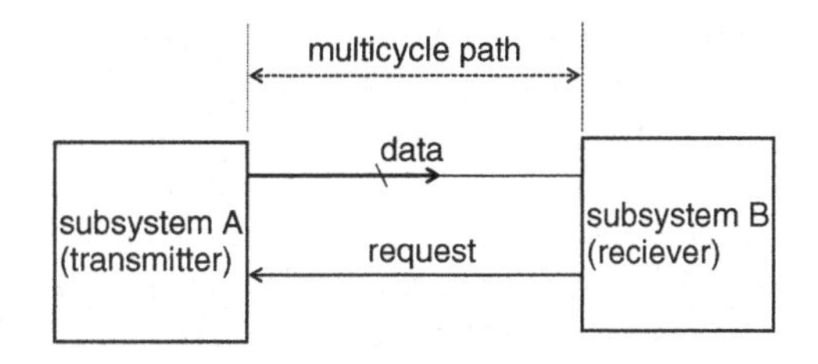

Fig. 1.11 Multicycle path at request driven data transfer point

This is a very simple depiction of handshaking protocol to transmit data and normally it involves more complex circuitry. Here it is simplified just to introduce the multicycle path concept. The RTL designer can specify this path as multicycle path in the SDC file.

Second type of multicycle path is that path between two points of the circuit which will take more than one cycle of clock due to its logic depth and hence will hamper the speed of the system. A simple example of this type of path would be a multi-level adder unit involving many operands. Figure 1.12 depicts this logic path.

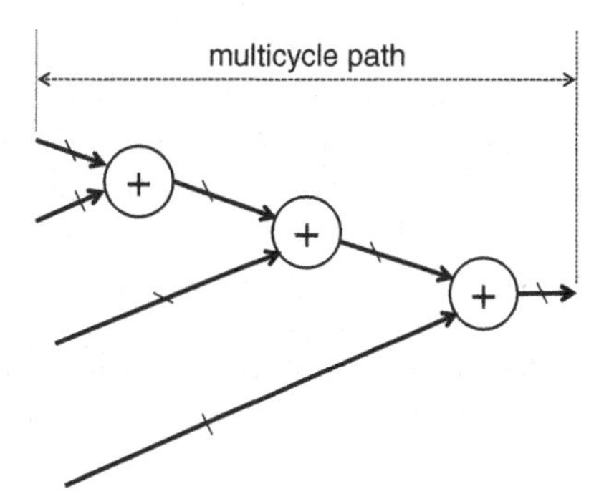

Fig. 1.12 Multicycle path due to multi-operand adder stage

As the output of this complex adder will be available only after few cycles of clock, the consumer of this output has to wait for that number of clock cycles. This hampers the speed of the system but still it is better for RTL designer to specify this multicycle path exception in the SDC file to help logic synthesis, STA and physical synthesis tools.

All the above has been discussed and explained well in detail with all other timing exception related fundamentals in Chapter 7.

1.6.7 Congestion Conscious RTL

As described earlier, after an RTL design is simulated, synthesized and verified for timing, functionality and test, it goes to backend phase. This includes placement of cells and routing of wires between the placed cells. After initial P&R, it is found that some of the areas in the chip will have such a high concentration of wires that the adjacent wires start interfering with each other, so much that functionality of the chip could be broken. This happens due to increased coupling capacitances between the adjacently running wires and also these congested areas become hot-spots because of higher power consumption in a small area. Naturally, this cannot be allowed and hence congested areas have to be decongested.

Decongesting at this late stage of design cycle, without going back to any of the previous steps, is not always possible. Eventually, RTL may have to be modified and all the steps after that have to be repeated. These iterations with RTL are very expensive and in most cases delay the delivery of the chip. So, even though alleviating congestion is a post layout job, an RTL designer should better write an RTL which is *congestion conscious* and then only hand it over to backend team.

Big question again is, how can an RTL designer write a *congestion conscious RTL*? An RTL designer should use an EDA tool which can give a good estimate of locations where congestion may be more severe. And, based on the output of the tool, RTL should be modified to decongest those areas. Congestion will be reported on those areas of the design where number of pins per unit estimated area of the design is high. For example, if a particular piece of code gets synthesized into a large macro having huge number of data lines as input and hence the area around this macro would show up as having a high congestion. What an RTL designer can do is to design the RTL or synthesis scripts in such a way that the design gets synthesized into smaller macros. This may mean higher die area consumed and the RTL design may need to get adjusted again to get the optimum *congestion* and *area* values. This kind of what-if analysis done by an RTL designer using an appropriate physical prototyping tool at RTL stage can help the backend engineers to close the design earlier.

All the above has been discussed and explained well in detail with all other congestion related fundamentals in Chapter 8.

1.7 Summary

RTL designer has a very challenging role. Not only he has to write an optimum code for desired functionality but also has to make sure that his RTL is friendly with all the later stages of the design flow. To achieve this he has to have a very good and deep understanding of all the concepts of these later stages. Surely, there are tools available to help him find the issues in his RTL but it is easy to get the best out of these tools if a designer has the knowledge of these concepts.

Cleaning and optimizing the RTL for all the later stages of the design flow, makes it a strong candidate to become a valuable IP (Intellectual Property) for the RTL designer and his company. The same IP can be reused in several other chips, if needed.

Please go ahead and read rest of the book to get a deep understanding of all the concepts introduced in this chapter.

Chapter 2
Ensuring RTL Intent

A user starts the design of his block, by describing the functionality of the block in the form of RTL. The RTL code is then synthesized – to realize the gate level connectivity that would provide the same functionality. Before synthesizing, the designer needs to ensure that the RTL actually implements the functionality that is desired. For this purpose, the designer runs a lot of simulations. The process of simulation shows that for a given set of input vectors, what would be the response. Only when the designer is sure of the RTL achieving the desired functionality, the RTL is sent for synthesis and subsequent steps. The main advantages of describing the design in RTL rather than logic gates are mainly two fold:

(i) Higher level of abstraction makes it easier to describe the design, compared to providing the gate level connectivity.
(ii) RTL simulations are much faster compared to the corresponding gate level representation. Hence, validation of the functionality can be much faster.

2.1 Need for Unambiguous Simulation

Since the RTL code has to ensure that it has achieved the desired functionality, the RTL code is subjected to rigorous simulations with many different vector-sets to cover all the functionalities of the device. These vectors are supposed to cover various situations – including:

- Normal mode of operation
- Specific corner-case situations
- Error/recovery handling etc.

Simulation is unambiguous till the time the RTL has only one possible interpretation. However some pieces of RTL code can have multiple interpretations. The simulator being used by the RTL designer will exhibit any one of the possible multiple interpretations. The designer is finally satisfied with the exhibited functionality. However, it is possible that during simulations, the interpretation that has been used

S. Churiwala, S. Garg, *Principles of VLSI RTL Design*,
DOI 10.1007/978-1-4419-9296-3_2, © Springer Science+Business Media, LLC 2011

by the simulator is different from what gets finally realized (synthesized). If this happens, even though, the design functionality is verified through simulation, the actual operation of the realized device does not exhibit the same behavior.

Functional verification of an RTL code (through simulation) is one of the most basic and very fundamental aspect of a chip design and therefore, it is imperative that any RTL code that is written is un-ambiguous in terms of functionality.

2.2 Simulation Race

The word *race* could have different meanings in different context. In the context of HDL simulation, the word *race* means same HDL code exhibiting two or more possible behaviors – both of which are correct as per the language interpretation. So, a *race* implies ambiguity. By its definition itself, *race* says that there are multiple interpretations, and, all of them are correct. A simulator is free to choose any one of these multiple interpretations. However, it is not necessary that the final realization will match the interpretation that your simulator chose.

Simulation race is usually encountered in *Verilog* code. *VHDL* does not have any simulation race but it has some other nuances – which we will see in Section 2.3.3. Also, *SystemVerilog* has provided additional constructs and semantics which provide additional information to the simulators so that their response is consistent. Thus, *SystemVerilog* has provided ways to do away with a lot of *race* situations. However, it is still possible to have a *SystemVerilog* code, where a portion still has a race. This can happen, if the designer has used *SystemVerilog* only at some portions of his code and not really exploited the full prowess of capabilities provided by *SystemVerilog* to avoid a race situation.

In case of a design having a race:

- Different simulators can give different results
- Different versions of the same simulator can give different results
- Same version of the same simulator can give different results – based on switches chosen or changes made to the code, which have seemingly no relation with the change in behavior being exhibited. For example, including some debug level switches or some debug type statements could change the result of the simulation

2.2.1 *Read-Write Race*

Read-Write race occurs when a signal is being read at the same time as being written-into.

2.2.1.1 Combinational Read-Write Race

Consider the following code-excerpt:

```
assign a = b & c;

always @ (b or d)
if (a)
        o = b ^ d;
```

In the above code-segment, *a* is written into through the ***assign*** statement – while being read through the ***always*** block. When *b* gets updated, it will trigger both the ***assign*** statements as well as the ***always*** block. The ***if*** condition in the ***always*** block will see the updated value of *a,* if the ***assign*** statement gets executed first. On the other hand, the ***if*** condition will see the old value of *a,* if the ***always*** block gets executed first. Depending upon which of the two behaviors is chosen by your simulator, you might get a different behavior. The above code segment is an example of a race.

Similar race-condition is also exhibited by the following code segment:

```
always @ (b or c)
a = b & c;

always @ (b or d)
if (b) o = a ^ d;
```

In the above code segment, *a* is being written into (through the first ***always*** block) – while it's also being read (through the second ***always*** block). Depending upon which ***always*** block gets triggered first, the second ***always*** block may see the updated value or the old value of *a.*

The solution to remove this ambiguity is very simple. In both the examples shown, the sensitivity list of the second ***always*** block should have *a* in it. This will ensure, as soon as *a* gets updated – the second ***always*** block gets triggered and values are re-evaluated with the updated value of *a.* In the worst case, the second ***always*** block might get triggered twice, but, the final values would be unambiguous. (Theoretically, this could still be a race, but, for all practical purposes – there is no ambiguity. See Appendix A for more details)

SystemVerilog provides a much more elegant solution to this kind of ambiguity. Instead of using the keyword, ***always***, you should use ***always_comb***. When you use ***always_comb***, there is no need to explicitly specify the sensitivity list. All the signals that are being read in this ***always_comb*** block will automatically be included in the sensitivity list.

2.2.1.2 Sequential Read-Write Race

Consider the following code-excerpt:

```
always @ (posedge clk)
b = c;

always @ (posedge clk)
a = b;
```

In the above code-segment, at the positive edge of *clk*, both the ***always*** blocks get triggered. So, *b* is being written into (through the first ***always*** block), while it is being read also (through the second ***always*** block). Assume that the first ***always*** block gets triggered before the second one. *b* gets updated. Then, the second ***always*** block gets triggered; *a* sees the updated value of *b*. Thus, in effect, the value of *c* percolates all the way to *a* through *b*. Consider another scenario. The second ***always*** block gets triggered before the first one. In this case, *a* sees the old value of *b*, and, then, *b* gets updated. So, depending on the sequence chosen by the simulator, the value on *a* could be either *c* (i.e. the new value of *b*), or, the old value of *b*.

The solution to deal with this kind of ambiguity is also very simple. For a sequential element, we should use a *Non-Blocking Assignment* (*NBA*). With an *NBA*, the Right-Hand-Side is read immediately, but, the updating of the Left-Hand-Side happens after all the reads scheduled for the current time have already taken place. Think of this as an infinitesimally small delay!!! We are considering this to be infinitesimally small, because in reality the simulation time does not move. It is just that all updates to Left-Hand-Side (of an *NBA*) happen after all the corresponding Right-Hand-Sides have been read. So, irrespective of what sequence is used for triggering the blocks, the events will occur in the following sequence:

(i) *b* and *a* will decide what values they should go to, but, they will not actually get updated. The relative sequence of the two ***always*** blocks is still undeterminable, but, that does not make any difference.

(ii) Subsequently, *b* and *a* will get updated. This updating happens after all read of Right-Hand-Side have already taken place. Since, *a* has already decided what value to go to (in the first step itself), so, any change in value of *b* is not going to impact the value of *a*.

Thus, the results can be made unambiguous through the use of NBA.

For the sake of correctness and completeness, it should be mentioned here, that with an NBA, after all the read has happened, and, then the left hand side gets updated; an updated LHS at this stage can trigger another sequence of reads. So, the updates are in effect taking place before some of the reads. However, from race perspective, this fact does not make any difference. The sequence of read and then update (which can then trigger more reads) remains unchanged.

2.2.2 Write-Write Race

Write-Write race occurs when multiple values are being written into a signal – at the same time.

Consider the following code-segment:

```
always @ (b or c)
if (b != c)        err_flag = 1;
else               err_flag = 0;
```

```
always @ (b or d)
if (b == d)        err_flag = 1;
else               err_flag = 0;
```

When *b* changes, both the ***always*** blocks get triggered. It is possible that in one of the ***always*** blocks, *err_flag* is supposed to take the value of *1*, while, in the other, it's supposed to take the value of *0*. The value that *err_flag* finally takes will depend on the sequence in which these ***always*** blocks get triggered. The one that triggers later – will determine the final value. Though *b* is the common signal in the two sensitivity lists, it has no role to play in creating the race. It is possible to have a similar race, with no signal being common in the two sensitivity lists. The race is being created because of the assignments in the two ***always*** blocks being made to the same signal.

always_comb (instead of the ***always***) as explained in Section 2.2.1.1 cannot be used here, because, same variable, *err_flag* cannot be assigned a value in two different ***always_comb*** blocks. Use of *NBA* as explained in Section 2.2.1.2 will also not solve this problem. The simplest solution to this ambiguity is to avoid updating a signal in more than one concurrent block. A signal should be updated in only one concurrent statement. In the following excerpt ***always_comb*** has been used just for convenience. For a combinatorial block, it is anyways a good practice to use ***always_comb***, rather than just ***always***.

Modifying the above code-segment – in line with the above guidelines, you get:

```
alway_comb
begin
if (b != c)        err_flag = 1;
else               err_flag = 0;
if (b == d)        err_flag = 1;
else               err_flag = 0;
end
```

The above code-excerpt is still wrong. This is not probably what one had intended. But, it does not have a race. It will behave in the same (incorrect manner) with all simulators!!! The following code-segment avoids the race as well as achieves the intended behavior:

```
alway_comb
begin
err_flag = 0;
if (b != c)          err_flag = 1;
if (b ==d)           err_flag = 1;

end
```

A Write-Write race can occur through combinations of:

- ***assign/assign*** statements – if two ***assign*** statements try to update the same variable. However, typically, these are caught easily in simulations – as the assigned variable might tend to go to an *X*.
- ***always/always*** statements – if two ***always*** blocks try to update the same variable. These situations might be encountered only in a testbench. When used in a design this will result in a Multiple Driver scenario after synthesis – which is a violation of basic electrical requirement. The simulator might not necessarily see the Multiple Driver scenario!!!

2.2.3 Always-Initial Race

Always-Initial race occurs when an ***initial*** block is updating a signal which also appears in the sensitivity list of an ***always*** block.

Consider the following code-segment:

```
initial
rst_n = 1'b0;

always @(posedge clk or negedge rst_n)
if (! rst_n)    q <= 1'b0;
else            q <= d;
```

Scenario 1

At the start of the simulation, the ***always*** block is triggered first. It now has to wait on a positive edge of *clk* or negative edge of *rst_n*. So, it now waits for the triggering event to happen. Then, the ***initial*** block is triggered. This causes *rst_n* to go to *1'b0* thereby causing a negative edge on *rst_n*. Since the ***always*** block was waiting for a negative edge of *rst_n*, this ***always*** block starts executing, taking *q* to *1'b0*.

Scenario 2

At the start of the simulation, the ***initial*** block is triggered first. It takes the *rst_n* signal to value *1'b0*. The negative edge on *rst_n* (in the ***initial*** block) has already happened, even before the ***always*** block could wake up. So, the negative edge on *rst_n* has been missed by the ***always*** block. So, *q* does not get initialized at the start of the simulation. The ***always*** block would wait for the next positive edge of *clk* or negative edge of *rst_n*.

Both the scenarios are valid as per *Verilog* language definition. Thus, a simulator might exhibit any of the above behaviors. These kinds of races are resolved through one of the following methods:

Before the advent of *SystemVerilog*, one method was to initialize using inactive values, and then, initialize after a while. So, the above code would be modified as:

```
initial
begin
rst_n = 1'b1;
#5 rst_n = 1'b0;
end

always @(posedge clk or negedge rst_n)
if (! rst_n)     q <= 1'b0;
else             q <= d;
```

The use of *#5* in the above example is indicative only. Important point is to put some delay – however small it may be. In the above example, at time *0*, the ***always*** block is not supposed to be executed. It would have just woken up (or, get armed – as some people say). So, irrespective of the order of execution, at time *0*, the ***always*** block gets armed, but, not executed. And, at time *5*, there is a negative edge on *rst_n*, which will cause the ***always*** block to be executed. The ***initial*** block in the example mentioned above is typically found in a testbench, rather than in a design. Synthesis ignores ***initial*** block as well as explicit delay specifications. Hence, such constructs are not used in design which will be later realized in terms of logic gates.

SystemVerilog provides a much more elegant method of removing the non-determinism (i.e. race). It says that all variables initialized at the time of declaration are initialized just before time *0*. *rst_n* can be initialized to *0* at the time of declaration itself (in the testbench). This means, at time *0*, there is no initialization, and hence, there will be no event. Thus, *a* deterministic behavior is forced by *SystemVerilog*.

2.2.4 Race Due to Inter-leaving of Assign with Procedural Block

Section 5.5 of the *IEEE Standard 1364–1995* (popularly called as *Verilog*) provides the following example code-excerpt:

```
assign p = q;

initial begin
q = 1;
#1 q = 0;
$display(p);
end
```

For this specific situation, different simulators are known to exhibit different behavior. When *q* is updated to *0*, the ***assign*** statement is supposed to reevaluate the value of *p*. Some simulators continue on with the current ***initial*** block to display *p*, before moving onto the ***assign*** statement. These simulators display a *1*. On the other hand some simulators suspend the current ***initial*** block, and, execute the ***assign*** statement. Then, they come back to the suspended ***initial*** block, and, display the new value of *p*, thus displaying a *0*.

The way to resolve this ambiguity would be to put a delay, before the ***$display***. This delay specification will force the execution to move to the ***assign*** statement, before executing the ***$display***. Alternately, use of ***$monitor*** instead of ***$display*** should show the final value of *p*. Appendix A discusses race implications due to interleaving of concurrent processes.

2.2.5 Avoiding Simulation Race

In his paper, "Nonblocking Assignments in Verilog Synthesis, Coding Styles That Kill!" Cliff Cummings of Sunburst Design recommends the following guidelines (reproduced verbatim) to avoid simulation race conditions in your Verilog RTL design:

(1) When modeling sequential logic, use nonblocking assignments.
(2) When modeling latches, use nonblocking assignments.
(3) When modeling combinational logic with an always block, use blocking assignments.
(4) When modeling both sequential and combinational logic within the same always block, use nonblocking assignments.
(5) Do not mix blocking and nonblocking assignments in the same always block.
(6) Do not make assignments to the same variable from more than one always block.
(7) Use $strobe to display values that been assigned using nonblocking assignments.
(8) Do not make assignments using #0 delays.

2.3 Feedthroughs

Feedthrough refers to a situation, where a signal, instead of just getting captured in the destination register, overshoots it and goes to the next stage also – within the same cycle, instead of waiting for one additional cycle. (The term Feedthrough has one more usage. We will look at another usage of the term in Section 8.4) Figure 2.1 shows an example, where – in each clock cycle, data from register *A* goes to *B*, and, that from *B* goes to *C*. Normally, a data which starts from Register *A* will reach *B* in one cycle, and, then, into *C* in yet one more cycle.

But, if the data from *A* reaches *B*, crosses it, and, goes to *C* – all within just one cycle, it is a situation of a *Feedthrough*. *Feedthroughs* can happen in VHDL also,

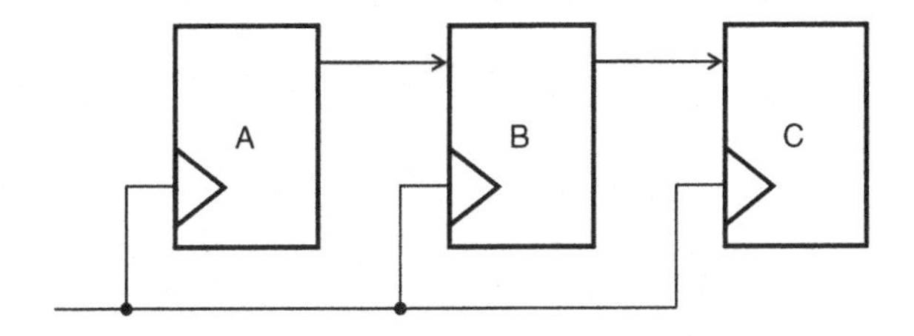

Fig. 2.1 Data transfer through registers

unlike *races* which happen only in Verilog. Some users refer to *Feedthrough* also as *race*. So, it should be understood that when spoken in the context of VHDL, a *race* typically means *Feedthrough*.

2.3.1 Feedthroughs Because of Races

One of the causes of *feedthrough* is races. Among multiple interpretations of a code, there might be one – which results in a *feedthrough*. Consider a Verilog code segment:

```
always @ (posedge clk)
q1 <= t1;

always @ (posedge div_clk)
q2 <= q1;

always @ (posedge clk)
div_clk <= ~div_clk;
```

Consider one possible scenario of events – for the above code. There is a positive edge on *clk*. *q1* will get updated. *div_clk* will also toggle. Both *q1* update and *div_clk* toggle will happen after infinitesimally small delay (of *clk* ***posedge***) – but at the same time stamp, because, they both have *NBA*. At the positive edge of *div_clk* (which happens after a while), the second ***always*** block gets triggered. Since this has been triggered after a while (actually, towards the end of the current time-stamp), *q1* could already have been updated. So, the updated value of *q1* will reach *q2*. Thus, *t1* has overshot *q1* and has gone into *q2*.

There is one more scenario (sequence of events) possible. This possible scenario is left up to you to work out. In that scenario, feedthrough would not happen.

One possible solution to this kind of situation is to avoid using *NBA* in clock-paths. In this case, the clock-path (generation of *div_clk*) has an NBA. If this NBA is replaced by the Blocking Assignment, the second ***always*** block will trigger instantaneously, before *q1* is updated. So, the correct way of writing the above functionality would be:

```
always @ (posedge clk)
q1 <= t1;

always @ (posedge div_clk)
q2 <= q1;

always @ (posedge clk)
div_clk = ~div_clk;
```

The example mentioned in Section 2.2.1.2 has two possible interpretations. One of those interpretations results in a *feedthrough.*

2.3.2 Feedthroughs Without Simulation Race

Feedthroughs can happen even if there is no simulation race. Consider the following *Verilog* code segment:

```
always @ (posedge clk)
begin
c = d;
b = c;
a = b;
end
```

The above code has no ambiguity, but, there is a feedthrough. *d* goes into *c*, and then into *b* and then into *a* – all within the same cycle. This is usually not, what was intended. The solution to this situation is simple. Use *NBA*!!!!

2.3.3 VHDL Feedthroughs

VHDL has a concept of *delta-delay*. The simulator assigns an infinitesimally small delay (called *delta*) whenever there is an assignment to a *signal* (assignments to *variables* are instantaneous). Consider a clock signal *Clk1*. *Clk1* is used as a master to derive another clock signal *Clk2*. *Clk2* is delayed from *Clk1* – by a few *deltas* (these, additional deltas could be due to clock-gating – for example). The following code-excerpt shows this situation:

```
process (Clk1)
begin
SIG1 <= Clk1; -- one delta from Clk1
Clk2 <= SIG1; -- Two deltas from Clk1
end process;
```

Clk1 is also used in the sensitivity list of a ***process*** block that updates a signal *q1*. There will be one delta – for generation of *q1*.

```
process (Clk1)
begin
if Clk1'event and Clk1='1' then
q1 <= data; -- one delta from Clk1
end if;
end process;
```

Clk2 is used in the sensitivity list of another ***process*** block (excerpt below) that samples *q1*. Because, *Clk2* is delayed by a few deltas, by the time the below mentioned ***process*** block is triggered; *q1* is already updated. So, the value of *q1* sampled in this ***process*** block would be the new-updated value.

```
process (Clk2) -- will trigger 2 deltas after Clk1
begin
if Clk2'event and Clk2='1' then
q2 <= q1; -- q1 is already updated!!
end if;
end process;
```

In this case, *data* will move to *q1* and then into *q2*, all within the same cycle of *Clk1*. So, there has been a *feedthrough*. You can avoid this, by ensuring balancing of delta-delays across the clock network. That means, on all the clock-paths, there would be exactly same number of deltas. So, *Clk1* and *Clk2* must have the same number of deltas. That will ensure that *q1* cannot be created before *Clk2*. And, if *Clk2* is created from *Clk1* – as in the above example, then, *Clk1* is not used directly to update *q1*. Rather, *Clk1* is delayed further by the required number of (in this case, 2) deltas – and that delayed version is used to update *q1*. However, there are three issues with this approach.

- Most of the designs today use Mixed-Language. And, in Mixed-Language, there is no Language Reference Manual (LRM) – to define the behavior. So, there are some differences in the behaviors of different simulators, with respect to counting of deltas – especially as a signal crosses VHDL/Verilog boundary. So, it is possible that what is balanced in one simulator is no longer balanced in another simulator.
- It is too much of a trouble to keep counting all these deltas along various clocks.
- This requirement of balancing of deltas along clock-networks is much more stringent than what's really needed. What's really needed is: Deltas along: Clk1 --> q1 --> reaching onto the sampling signal have to be more than the deltas along Clk1 --> Clk2. So, if there is a minor difference in the delta-count along two clock-paths, but, there are enough delta-counts on the data-line, then, the clock network is as good as balanced.

Alternately, you may put a small delay while assigning a value to a signal, which will be used subsequently in another ***process***. The concept is shown in the following example code-snippet:

```
process (Clk1)
begin
if Clk1'event and Clk1='1' then
q1 <= data AFTER 1 NS;
end if;
end process;
```

This explicit ***AFTER*** clause ensures that *q1* is updated only after its older value is sampled in another ***process*** block. Effectively, this ***AFTER*** clause is acting as if it's creating a very high number of deltas in the data-path. This solution has its own risk, that, if the circuit's functionality was dependent on this specific value of explicit delay, the circuit might finally fail; because, synthesis will not honor this explicit delay assignment.

In fact, some designers always want to avoid explicit delay assignments – so that they inadvertently do not create a delay-dependent functionality. Besides, using explicit delays could cause the simulation to slow down considerably. On the other hand, some designers always want to use explicit delay assignments – so that they don't get into feedthrough conditions – due to mismatch of deltas.

2.4 Simulation-Synthesis Mismatch

Now that you have written your RTL in a manner, that there is no scope for ambiguity, you can simulate your design to ensure that it achieves the functionality that you desire. However, you have to ensure that not only the RTL should show the functionality that you desire, but, even the gate-level netlist that you will obtain from it after synthesis will also continue to exhibit the same functionality. Otherwise, there is no use, if RTL exhibits one functionality, but, the realized gate level shows a different functionality.

Simulation-Synthesis Mismatch refers to a situation, wherein, a given RTL showed some simulation results; however, when the same RTL was synthesized, and, the realized gate-level netlist was simulated using the same vectors, it exhibited a different behavior. Some of the most common reasons for Simulation-Synthesis Mismatch include:

- *Races* (as explained in Section 2.2)
- *Explicit Timing in RTL*. Say, the RTL code had an explicit delay assignment. When synthesis is done, the functionality is realized using technology gates. The delay for these technology gates could be totally different from what was

specified in the input RTL. And, if the functionality was dependent on the delay values, then, it is possible that the same functionality might no longer be visible – after synthesis, as the delay values have now changed.

- *Missing sensitivity list.* The RTL code-segment below is for a combinatorial block that misses some signals in the sensitivity list:

```
always @ (a or b)
if(sel)    z = a;
else       z = b;
```

Here, while simulating the RTL, if *sel* was changed, the **always** block would not be triggered, and, hence, *z* will not exhibit the new value. But, after synthesis, this above code will become a MUX. And, when the gate-level circuit is simulated, as soon as *sel* changes, the value of *z* will be updated. Thus, the simulation results on gate-level design could be different from what was seen from RTL simulation. Use of *SystemVerilog* construct ***always_comb***. It is again helpful here – as there will be no need to explicitly specify the sensitivity list.

- *Delta unbalancing.* It is possible that the RTL was written using balancing of Deltas (as explained in Section 2.3.3). However, during synthesis, the tool did its own adjustments in terms of inserting some additional buffers (to meet the load requirements), or, removing certain gates (to improve the timing). These adjustments could modify the delta-balancing. Usually, this is not encountered too often, because, most designers use Verilog for gate-level and this issue of delta-balancing does not come into picture.
- *Initial Block.* Synthesis simply ignores the ***initial*** block. So, if an ***initial*** block is used to achieve a desired functionality at RTL stage, then, the realized gate-level netlist will not exhibit the same functionality. Usually, ***initial*** block should be a part of the testbench – used to initialize the circuit through external inputs, rather than being a part of the design itself. This same testbench will then be able to initialize the gate-level circuit also.
- *Dependency on X*: *x* or *X* is something that is available only in modeling. In the actual hardware, there is no such thing as *x*. So, if a dont_care value is being assigned or checked for in the RTL, the same behavior might not be visible in the synthesized gate-level netlist.
- *Comparison with Unknown*: A comparison with *x* or *z* could result in simulation-synthesis mismatch. *x* does not have any counterpart in the physical world. *z* means tri-state, but, usually, in the physical world, there will be some value – available on the net. So, comparison with tri-state would not have any meaning in the physical world. Similarly, if a design is dependent on use of === or *!==*, or ***casex*** or ***casez***, there could be a mismatch, because, these comparisons consider *x* and *z* as if those were also a value. In the world of *VHDL*, *U* or *W* related comparisons could also result in simulation-synthesis mismatch, because they also don't have a physical-world equivalent that could be realized using logic gates.

- *Careless use of variables*: In *VHDL* RTL code, ***variables*** are not allowed by language to cross ***process*** boundaries. Consider the following code segment:

```
process(rst_n, clk)
variable v1_var : std_logic;
begin
v1_var := '0';
if (rst_n = '0') then
q <= '0';
elsif (clk'event and clk = '1') then
q <= data;
v1_var := '1'; -- simulator will respect this, but, synthesis will ignore
end if;
sig1 <= v1_var; -- sig1 will show simulation/synthesis mismatch
end process;
```

The variable *v1_var* stays within the ***process***, but, goes outside the "*clock*"ed process. Synthesis will ignore the second assignment to *v1_var*, while, simulation will assign a value of *1* to *v1_var*. This can also result in simulation-synthesis mismatch. This mismatch will then be propagated to *sig1*, which is reading the value of *v1_var*. Reusability Methodology Manual (written jointly by *Mentor Graphics* and *Synopsys*) recommends against using ***variables***, because of their potential to cause such simulation-synthesis mismatches.

As an RTL designer, you need to avoid all the above mentioned situations and constructs – which can result in simulation-synthesis mismatch.

2.5 Latch Inference

Synthesis tools infer a latch, when a register is updated in some branches of a procedural block, but, not in all branches of the block. There might be instances of very large and complex procedural blocks, with a huge number of branching. Due to some omission, it is possible that for a specific register, it misses an assignment in some of the branches. This will result in synthesis tool inferring a latch. The following code excerpt will infer a latch:

```
always @ (a or b or c or d or e)
begin
if (a)
    begin
    if (d)    r = 1'b0;
    else      r = 1'b1;
    end
else
```

```
if (b)
begin
        if (e) r = 1'b0;
end
else
        if (c) r = 1'b1;
end
```

In the above code segment, though, not apparently visible – there are two branches, where, the value of *r* is not updated:

(i) *a* is not true; *b* is true; and *e* is not true.
(ii) *a*, *b*, and *c* are each individually false.

Because of missing assignment to *r* in these two branches, latches will be inferred. Latches can create complications in DFT (explained in Chapter 6). Hence, most RTL designers want to avoid latches (unless, the design is supposed to be latch based). Thus, you need to ensure that your RTL code does not infer a latch unintentionally.

SystemVerilog allows the use of keyword ***always_comb*** (instead of ***always***) to denote intent of combinatorial logic. And, if you want to infer a latch, you should use the keyword ***always_latch***. Software tools are allowed to flag a warning, if they find that the logic being inferred does not match the intent specified through ***always_comb*** or ***always_latch***. Thus, use of ***always_comb*** can warn you against unintentional latch inference.

Latches can also be inferred through ***case*** statements. Consider a 2-bit signal *sel* – which can take 3 values, viz: *00*, *01* or *10*. So, the case statement is written as:

```
case (sel)
2'b00: out = data1;
2;b01: out = data2;
2'b10: out = data3;
endcase
```

A synthesis tool might not be aware that *sel* cannot take the value *2'b11*. So, it will create a latch for the branch (*sel=2'b11*) – for which *out* has not been assigned any value. Synthesis tools allow pragmas embedded in the RTL (such as: *full_case*) to let the synthesis tool know – that all the possible values have been specified. This pragma tells the synthesis tool not to infer a latch for missing branches. However, if you miss a branch by mistake, you would not get any indication.

SystemVerilog has introduced a keyword – ***priority***, which serves multiple purposes. It is not a pragma; rather, it is a part of the language. So, this is understood not just by synthesis tools, but, also by the simulators and formal tools. And, if any tool sees that the case-selection variable takes a value that is not specified, it will give an Error. Consider the following code segment with the use of ***priority*** keyword:

```
priority case (sel)
2'b00: out = data1;
2;b01: out = data2;
2'b10: out = data3;
endcase
```

In this case, synthesis tool will not infer a latch. And, during simulation, if *sel* is found to take a value of *2'b11*, the simulator will give an Error. Just the presence of ***priority case*** does not mean complete protection against unintentional latch inference. ***priority case*** only means all the possible branches are specified. It is still your responsibility to provide the values to all the variables in all the branches. Consider the following code-excerpt:

```
priority case (sel)
2'b00: {out1, out2} = {data1, data2};
2;b01: {out1, out2} = {data3, data4};;
2'b10: out1 = data5;
endcase
```

The ***priority*** keyword only tells that all the possible branches have been specified. But, *out2* has not been updated in one branch. So, a latch would still be inferred for *out2* – despite using the ***priority*** keyword.

2.6 Synchronous Reset

Consider a flop *q* which can be cleared synchronously through assertion of signal *rst_n* (active-Low). This can be modeled in many ways. Some of them are given below:

```
always @(posedge clk)
if (!rst_n)
   q <= 1'b0;
else
   q <= d;
```

Or:

```
always @(posedge clk)
if (!d)
   q <= 1'b0;
else
   q <= rst_n;
```

Or:

```
always @(posedge clk)
q <= d & rst_n;
```

Or:

```
assign temp = d & rst_n;

always @(posedge clk)
q <= temp;
```

Or:

```
always @(posedge clk)
case (rst_n)
   1'b0: q <= 1'b0;
   1'b1:  q <= d;
endcase
```

All the various code-excerpts given above will behave the same – in terms of functional simulation. For the given example code-excerpts, even the synthesized netlist will have the same circuit realization as shown in Fig. 2.2.

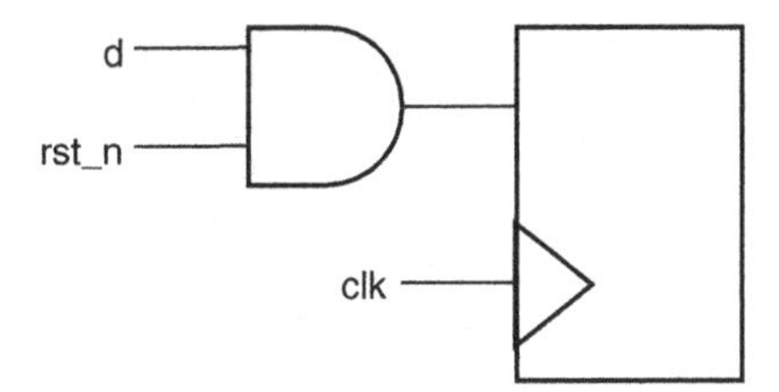

Fig. 2.2 Synchronous reset

As can be seen from some of the code-excerpts as well as the circuit realization, *d* and *rst_n* are interchangeable. If these signals were named something less meaningful (say: *a* and *b*), there is no way for a tool (or, even a human being) to distinguish between data and synchronous reset – atleast in some of the styles. Thus, it might appear that there is no need to distinguish amongst data and synchronous reset – because simulation results as well as the synthesized circuit are anyways the same. In terms of simulation results, there is actually no need to distinguish between data and a synchronous reset. However, synthesis tools try to treat data and synchronous reset signals slightly differently.

Certain ASIC libraries mark a synchronous reset pin through an attribute. Synthesis tool would try to connect the synchronous reset signal of the design to such pins. In the absence of such a pin, the synthesis tool uses data pin itself to achieve the synchronous reset functionality. But, it still attempts to put reset as close to the flop as possible. This is because, reset being one of the control signals – synthesis tools try to have minimal combinatorial logic on it. The presence of synchronous reset is reported as part of register inference. For Synopsys® synthesis tool, the report would look something like:

Register Name	*Type*	*Width*	*Bus*	*MB*	*AR*	*AS*	***SR***	*SS*	*ST*
q_reg	*Flip-flop*	*1*	*N*	*N*	*N*	*N*	***Y***	*N*	*N*

However, if the synthesis tool is unable to identify a signal to be synchronous reset, it might not be able to give the differential treatment to synchronous reset, and treat those signals at par with data. If synchronous reset is not detected, the report would look like:

Register Name	*Type*	*Width*	*Bus*	*MB*	*AR*	*AS*	***SR***	*SS*	*ST*
q_reg	*Flip-flop*	*1*	*N*	*N*	*N*	*N*	***N***	*N*	*N*

Because of this reason, synthesis tools should be indicated which signal acts as a synchronous reset. Consider the following code-excerpt:

```
always @(posedge clk)
if (!rst_n)
   q <= 1'b0;
else
   q <= data1 & data2;
```

Figures 2.3a and 2.3b show two possible realizations of the above functionality.

While both the circuits are functionally equivalent, a synthesis tool should prefer to realize the circuit as shown in Figure 2.3a, where, the synchronous reset is closer to the flop. However, for this, the synthesis tool has to know, which signal is synchronous reset. This can be conveyed through a pragma (*sync_set_reset* for

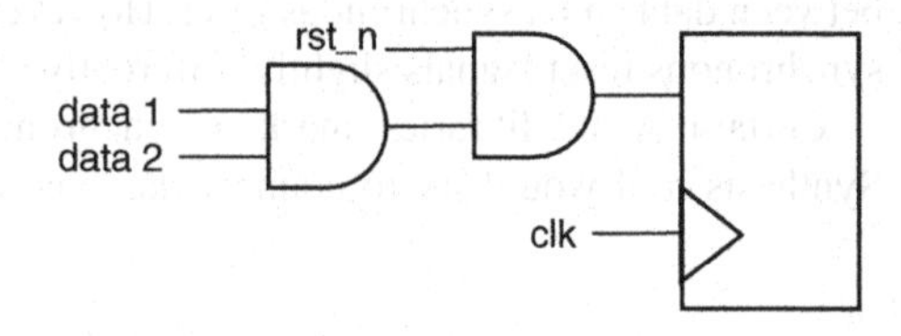

Fig. 2.3a Preferred realization

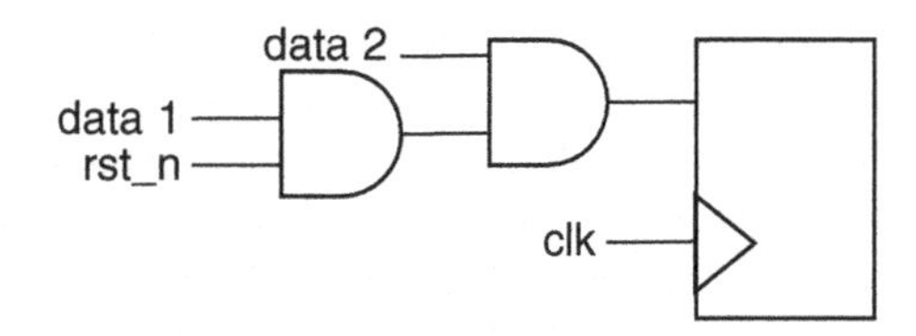

Fig. 2.3b Alternative realization

Synopsys synthesis tool) embedded in the RTL. Even with the pragma embedded in the RTL, the synthesis tool still depends upon the RTL structure to identify which signal is reset and which one is preset. For proper recognition of synchronous reset signals, it is important to have both the pragma as well as the right structure of the RTL.

In order to clearly communicate the intent of synchronous reset both for human understanding as well as synthesis inference, it is always best to code synchronous reset, using the style shown in the code excerpt below:

```
//pragma sync_set_reset rst_n
always @(posedge clk)
if (!rst_n)
   q <= 1'b0;
else
   q <= d;
```

This results in the following inference report:

Register Name	*Type*	*Width*	*Bus*	*MB*	*AR*	*AS*	***SR***	*SS*	*ST*
q_reg	*Flip-flop*	*1*	*N*	*N*	*N*	*N*	***Y***	*N*	*N*

If we remove the *sync_set_reset pragma* from the above code, the inference report would get changed to:

Register Name	*Type*	*Width*	*Bus*	*MB*	*AR*	*AS*	***SR***	*SS*	*ST*
q_reg	*Flip-flop*	*1*	*N*	*N*	*N*	*N*	***N***	*N*	*N*

Within the synchronous (i.e. clocked) portion of a sequential block, the first level *if* has to be the synchronous reset and then the entire data logic comes in the *else* section. This ***if*** condition should be for the asserted state of the synchronous reset signal.

The discussion mentioned in this section applies equally to synchronous preset also. In case both reset and preset are present, the first ***if*** condition has to be for the dominant of the two. The other synchronous signal should appear in the ***else if*** condition. And, the data logic should appear in the ***else*** condition. The code-excerpt below shows an example with both asynchronous and synchronous preset and reset. Both the signals are active Low and preset dominates over reset. In most cases, not all of these branches need to be present.

```
always @(posedge clk or negedge async_reset_n or negedge
async_preset_n)
  if (!asynch_preset_n) // Asynchronous preset – most dominant
    q <= 1'b1;
  else if (!async_rst_n) // Asynchronous reset
    q <= 1'b0;
  else begin // start of synchronous/clocked portion
    if (!synch_preset_n) // Synchronous preset – dominant
      q <= 1'b1;
    else if (!sync_rst_n) // Synchronous reset
      q <= 1'b0;
    else // data assignment
      q <= d;
  end
```

2.7 Limitations of Simulation

Simulation is the most popular and one of the most reliable methods used for validating that the HDL code meets the desired functionality. However, you should understand the major limitations of simulation. Some of the most important limitations are:

- As explained in Section 2.2, sometimes the same RTL code can be interpreted in multiple ways by a simulator. So, if your code is written in a manner that can give multiple results, your simulator will pick any one of those interpretations. Your simulation might pass with that interpretation, while the actual functionality realized could be different.
- Simulation based verification's effectiveness is limited to the quality of vectors and the quality of monitors being applied. With designs being so huge and with many storage elements, it is simply impossible to exercise the design for all possible situations. Vectors decide how the various parts of the circuits are being exercised. So, if a faulty portion of the circuit is not even being exercised, simulation will not be able to catch the fault. Monitors refer to what are you observing or checking for during the simulation. So, even if a faulty portion has been exercised, unless you are checking for something that depends on the values on that portion of the circuit, you will not be able to detect the fault.

- Even though, simulators have a concept of timing, they are not self-reliant at timing based verification. At the RTL level, most designs don't have timing. Even if there is some timing mentioned, it is not necessary that the same timing would be realized after the design is synthesized. So, RTL simulation does not give any sense of timing – other than being just accurate at the level of clock-cycle. And, at gate level, HDL languages do not provide for a good timing model. The best that they provide is *specify* block for Verilog and *VITAL* (*VHDL Initiative for Timing in ASIC Libraries*) for VHDL. These mechanisms are mostly place-holders for actual timing numbers. They don't do any delay calculation themselves. In order to do accurate timing simulation on a gate-level design, you have to do the delay calculation outside simulation, and bring back the delay values to the simulator through SDF back-annotation. This concept is discussed in the next chapter. So, for accurate timing simulation there is a dependence on an external delay calculation mechanism.
- Besides functionality, there are a lot of other aspects (test, power, routing congestion, etc.), which are important for a design. Simulation does not do anything to validate any of these aspects of the design. Though with the advent of some new formats (CPF/UPF) to specify power related intent, simulators can now do some validation of power aspects.

Because of these limitations, it's now a standard practice to use static rule checkers as part of the verification flow. These tools will verify many aspects of the design without the need for accurate vectors. These static checkers could be assertion based formal tools or rule-based checkers. In both cases, significant errors can be found that would otherwise go undetected.

Chapter 3
Timing Analysis

Timing Analysis of an ASIC design used to be traditionally done through *simulation*. The process involved applying a set of vectors and checking if the various signals are available at the desired time – at various points in the design. However, this process was too much dependent on the designer's coverage of the test-vectors. Hence, there was always a risk of missing some vector– which will actually not meet the timing and can result in failure to achieve the desired frequency. With increasing chip complexities, it became almost impossible to ensuring a complete and exhaustive coverage of vectors.

Around mid-90s, another concept of timing analysis started becoming popular. This is called, *Static Timing Analysis* (*STA*). In STA, there are no vectors applied. The process of STA computes the range of timing within which the signals will be available at various points in the design and compares them against the required time. *STA* refers to analyzing the timing aspect of the circuit – to ensure that the circuit can operate reliably at the desired frequency of operation. The concept of *STA* became quite popular around mid-90s. Before that, designers were more used to *Dynamic Timing Analysis*.

Inspite of STA being the practice, some designers still take their design through *Full Timing Gate Level Simulation* (*FTGS*), in order to do additional validation of timing – besides doing the STA. Instead of focusing on the behavior of specific timing tools, this chapter explains the fundamental concepts of timing, and, will also compare and contrast STA against Dynamic Timing Analysis – on some important aspects. Once the basics of timing are well understood, understanding the behavior of any specific timing tool would be much simpler.

3.1 Scope of STA

The scope of STA is limited to only validate the ability to meet the desired timing goals. STA does not do anything to deal with the logical functionality of the design.

S. Churiwala, S. Garg, *Principles of VLSI RTL Design*,
DOI 10.1007/978-1-4419-9296-3_3, © Springer Science+Business Media, LLC 2011

3.1.1 Simulation Limitations in the Context of Timing

The design's functionality is validated using simulation of the RTL. After *synthesis*, the functionality of the design's gate-level realization can also be validated through simulation, or, through establishing a formal equivalence of the synthesized netlist with respect to the validated RTL. However, the RTL does not have detailed timing information – in terms of delays through various paths etc. The concept closest to timing that it understands is *cycle-accurate. Cycle-accurate* means it would be known which signal will reach where in how many clock cycles. But, exactly when in that clock cycle is still not known. Besides, concept of cycles can only be applied on portions of the design which are synchronous.

Besides, the cycle-based behavior is simply assumed, rather than really validated by the simulator. Assume that the following RTL code excerpt is taken through RTL simulation. The simulation will simply assume that during the next cycle, bussed-signals *a* and *b* will reach the register bank *S*.

```
always @(posedge clock)
   S <= a * b;
```

The simulator will not validate this assumption, even if the clock frequency is increased drastically, or, even if the width of the busses *a* and *b* are increased significantly, or, even if there is more complex logic in the data-path. At the RTL stage, the simulator will simply assume that by the next edge of clock, the value of the expression (on the right hand side) would be evaluated and available!! So, after the RTL simulation has passed, you do know that the design seems[1] to be correct as per the desired functionality. However, you do not yet know – whether the design can exhibit the same desired functionality at the required frequency.

Alternately, you could do the simulation at the gate-level. However, gate level simulators (such as *Verilog* or *VHDL* based) do not have a very good timing model. Their models do not vary the delays based on several important factors on which the delays actually vary on silicon. Hence, even gate level simulation is not very reliable for predicting the design's ability to meet the timing. *SDF Back-annotation* described later in this chapter explains a mechanism – through which gate level simulation can have much better and accurate delay measurement through simulation. However, that mechanism itself depends on timing analysis. Alternately, much more accurate analog simulations may be used. But, their runtime is too high.

Even if you could solve some of these issues, the fundamental issue with simulation is ensuring exhaustiveness of all the applied vectors. So, it is possible that a super-accurate analog simulation (such as *SPICE*) shows the circuit to be good

[1] Note, the use of the phrase, "seems to be correct". This is because, in simulation, you are validating the design against the vectors that have been applied, but, there is no guarantee that the vectors have been applied for all possible cases. Hence, there is a possibility that the design might fail for a situation – for which the vectors were not applied.

enough to operate at the desired frequency; however, the design is actually subjected to a set of vectors against which the device was not simulated. And, for this set of vectors, the design is not able to meet the timing.

3.1.2 Exhaustiveness of STA

This is the fundamental reason for the sharp increase in the popularity of STA. STA computes both the minimum as well as the maximum range of delay values for all possible combinations of inputs. So, it does not depend upon a designer to specify the exhaustive combination of vectors. If the design is STA clean, it pretty much means that the design will meet the specified timing intent of the designer. The only reason why the design could still fail timing could just be simply some bug in the STA tool used, or, more importantly, incorrect specification of the intent by the designer. In fact, a lot of importance needs to be given to ensure that the timing intent specified (by the designer) during STA is correct.

At the RTL, there is no STA – because, the circuit realization itself is not available. The first stage, when STA can be done is during gate level netlist. After that, as design proceeds further down the implementation phases, the STA might be performed multiple times with increasing accuracy, as, interconnect delays, cell delays, clock path delays etc. get more and more refined progressively. Before Place and Route, since interconnect delays are not yet available, hence, an estimate of those are taken.

3.2 Timing Parameters for Digital Logic

On real CMOS devices, the circuits actually exhibit analog behavior. Digital behavior is an abstraction (or, modeling) technique – which makes the analysis simpler. The sections below explain the digital abstraction for delay and slew – which are actually characteristic of analog behavior. This digital abstraction is put in a set of files called *library*. This library is given as one of the inputs to the tools that need access to the timing parameters; and the tools pick up the data of interest as and when needed.

3.2.1 Delay Parameters

Delay is defined as the time between the application of an input and the observation of the response at the output. Let us consider an *AND* gate. If its input *A* is kept at *1*, and a transition is applied at *B*, a similar transition would be visible at the output *Z*. Say, if the transition occurs at *B* at time *t1*, and, the corresponding transition occurs at *Z* at time *t2*, then, the delay for the *B* to *Z* path is said to be *(t2 – t1)*. However, in

Fig. 3.1 Delay measurement

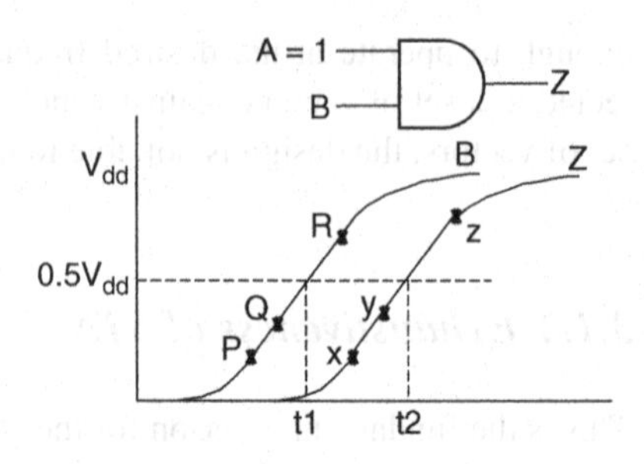

reality, the transitions at *B* or *Z* do not happen instantaneously. Figure 3.1 shows the corresponding input and output waveforms.

For the purpose of delay measurement, what should be *t1*? Should it be at point *P* or *Q* or *R* or some other point? Similarly, for the measurement of *t2*, which point should be chosen? Should it be x, y or z? Depending upon the points chosen for the measurement, the delay value for the *B* to *Z* path through the *AND* gate would be different.

There is no specific point which has to be chosen as the reference. Usually, it should be something where the signal transition shows a linear variation against time. For CMOS designs, this happens during the middle part of the transitions. The initial and the trailing portions of the transition are typically not linear; hence, these should not be used as the reference points. Hence, just the delay value of the *AND* gate is not sufficient. The delay specification has to include the point at which measurements started, and, the point at which the measurement was completed. Usually, these points are expressed as a percentage (or, fraction of V_{dd}).

In the example Fig. 3.1, $0.5V_{dd}$ has been chosen as the point of measurement for both *t1* as well as *t2*. It is not necessary that both *t1* and *t2* should have the same fraction. One could decide to use $0.5V_{dd}$ for start measurement (*t1*) and $0.4V_{dd}$ for end measurement (*t2*). Similarly, the measurement points could be different for rising waveforms and falling waveforms. The excerpt below shows an example of the delay measurement point specification in a hypothetical timing library (in SLF format)

```
input_threshold_pct_fall      : 45 ;
output_threshold_pct_fall     : 45 ;
input_threshold_pct_rise      : 45 ;
output_threshold_pct_rise     : 45 ;
```

So, when a transition is applied at the input, the output would show the corresponding response after the corresponding delay. Suppose, there is a glitch at the input. So, the input signal has a transition. The output is supposed to make a transition. But, before the output can switch to its intended value, the input comes back to its original level. In such a case, the output might not change its state. Effectively, the glitch at the input has not reflected at the output. Or, the cell has eaten up the glitch. This behavior of cell delay is called *inertial delay*. Another behavior is *transport delay*, where, the glitches at the input are transmitted as is at the output. Transport delay behavior is exhibited by wires.

3.2.2 Slew Parameters for Digital Logic

Slew (also called as *transition time* or *ramp time*) refers to the time that it will take for the signal to go from *0* to *1* (and, vice-versa). Let us say, a transition starts at time *t1* and is completed at time *t2*. So, the time *(t2 – t1)* is the *transition time*. Consider Fig. 3.2, which shows a transition.

Fig. 3.2 Slew measurement

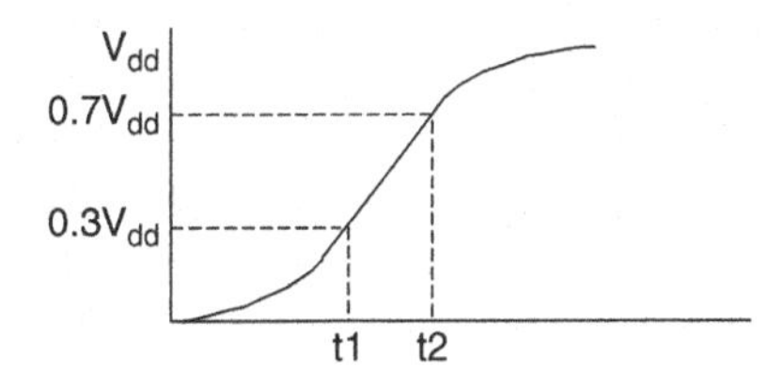

Now, once again, the issue of the starting point and end point for the measurement is applicable. There are no standard measurement points. Hence, the slew specification should include the points at which measurement was started and at which the measurement was stopped. In the given example, *t1* was measured at $0.3V_{dd}$ and *t2* was measured at $0.7V_{dd}$. Some of the more commonly used combinations are: $0.1V_{dd} - 0.9V_{dd}$; $0.2V_{dd} - 0.8V_{dd}$: $0.3V_{dd} - 0.5V_{dd}$; $0.3V_{dd} - 0.7V_{dd}$. Among these, $0.1V_{dd} - 0.9V_{dd}$ pair is not used – especially with reduced geometries. However, if you refer to very old libraries, it is possible that this pair might have been used. The excerpt below shows an example of the slew measurement point specification in a hypothetical timing library:

slew_lower_threshold_pct_rise : 35 ;
slew_lower_threshold_pct_fall : 35 ;
slew_upper_threshold_pct_rise : 65 ;
slew_upper_threshold_pct_fall : 65 ;

So, if you are comparing the speed of two different libraries or two different data sheets, don't just look at the delay or slew numbers. Also, look at the points at which the measurements were made. A gate with an output slew of *0.3* ns measured at $0.1V_{dd} - 0.9V_{dd}$ is transitioning faster than another gate with an output slew of *0.2* ns, but, measured at $0.3V_{dd} - 0.5V_{dd}$!!!

3.2.3 Delay and Slew Measurement

Delays and *slew* values are available in the library. So, STA engines can obtain the values from the libraries. However, for putting the values in the library, the values are obtained through analog simulation (*SPICE* based). So, a characterization engineer will measure the *delay* and the *slew* against a lot of parameters which effect delay and slew, and, put the corresponding values in the library.

For an output pin, the *slew* is measured by the characterization engineer using analog simulation and put into the table. For an input pin, the *slew* is just a specification, rather than a measurement. For example, for an *AND* gate, the characterization engineer will do a set of analog simulations with various conditions, and, will put for each of those conditions, what will be the delay and the slew at the output of the *AND* gate. The characterization engineer will also put the conditions for each of those delays/slews. These conditions also include the points at which the measurements were made – as explained in previous sections.

During *STA*, the applicable conditions would be determined, and the delay values corresponding to the conditions would be determined from the library. Similarly, the slew would be determined – at the output of the gate. This slew at the output of the gate forms the basis of slew computation at the input of the next gate. This computation of slew at the input of the next gate is based on the way the signal moves through the interconnect wire. Figure 3.3 explains the concept pictorially.

The loop represented by $A \rightarrow B \rightarrow C \rightarrow D$ is evaluated many times, till the delay and the slew computation etc. is done for all the points of interest. The box bounded

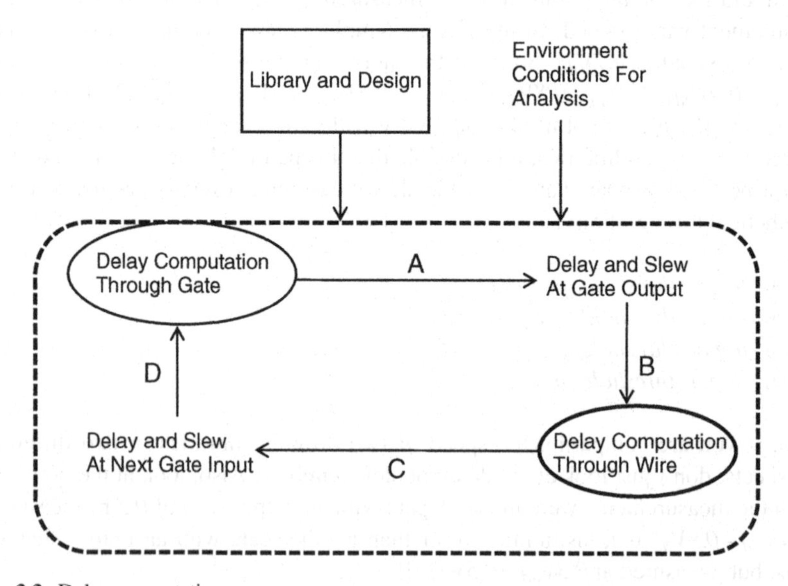

Fig. 3.3 Delay computation

by broken lines represents the core of an STA engine. In addition, the STA engine also validates that with the computed timing numbers – is the timing intent met or not? For a circuit having loops, a break condition has to be modeled; else, the above computation would keep on going forever. The output of a gate will come back at its own input and thereby – triggering the computation of its output once again. This is discussed later in Chapter 7.

Thus, the final computations depend on the exact network, the library chosen, the environment conditions chosen for analysis etc. Environment conditions indicate factors which can impact delay or slew, e.g. input transition, temperature, etc. Also, it is worth noting that slew at any one point impacts the delays at nodes further down in the circuit. Hence, it is important that the slew does not become very large at any node in the design.

3.3 Factors Affecting Delay and Slew

Delay and slew through a gate are impacted by many discrete and continuous factors. Some of the important factors are:

- Gate's Geometry and Schematic
- Specific Path
- Specific Directions of the Transitions
- Conditions on Other Pins
- Load on the Gate
- Input Slew
- Temperature
- Voltage
- Fabrication Process

3.3.1 Discrete Factors

These are factors which don't have a range of continuous values. These factors can have few specific values or points.

3.3.1.1 Gate's Geometry and Schematic

The *geometry* of a gate refers to the physical dimensions of the transistors used to form the gate. *Geometry* is usually a function of the specific technology. For example, 90 nm is a different *geometry*, compared to 45 nm. *Geometry* has a very significant role in deciding the delay through the gate. Lower *geometry* means faster gate (i.e. smaller delay). *Schematic* refers to the interconnection of the transistors – within the gate. So, a *NAND* gate will have lesser delay, compared to an *AND* gate, because, *AND* gate has an extra stage, compared to the *NAND* gate.

The geometry is represented as a property of the *library* itself. For example, a 32 nm library would be different from a 90 nm library. Within a specific library, gates with different functionality have different schematic – and hence, are named differently. A specific gate has a specific schematic of its own. Schematic is a combination of how various transistors are connected, and, their relative sizes. The interconnection of the transistors decides the functionality of the gate. The relative transistor sizes decide the electrical characteristics of the gate, such as drive strength, input load etc. For the same functionality, several different variations in gates are usually available in the library – in order to allow trade-off among power, size and performance. The different variations (but same functionality) are also represented in the form of different gates within the library. Gates with same functionality but different electrical characteristics are also considered unique and have different schematics. So, when you talk about *AND2B* gate in a specific library, you have fixed the schematic (specific interconnect of transistors to decide the functionality) and the geometry (specific sizing of transistors to decide the electrical characteristics).

Thus, each gate would have a different value for the delay, slew and other electrical characteristics. So, for specific performance requirements, you need to choose the appropriate geometry. And, within that geometry, the specific gate would tell you – what kind of performance to expect from that specific device.

3.3.1.2 Specific Path

Different paths of a gate have different delays. Figure 3.4 shows the schematic of a CMOS *NAND* gate.

Fig. 3.4 CMOS NAND gate

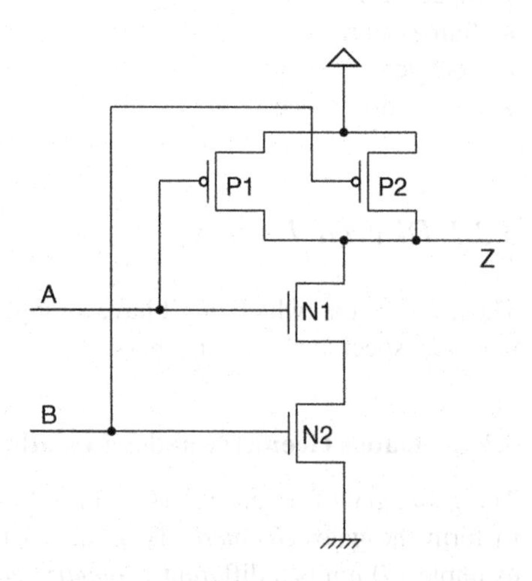

For this *NAND* gate, if the output goes high due to a transition on *A*, the current flows through transistor *P1*. Similarly, when the output goes high due to a transition on *B*, the current flows through transistor *P2*. Thus, the delay through *A* to *Z* path will be different compared to the delay through *B* to *Z* path – because different circuit elements are involved. So, for deciding the delay or slew through a gate, you need to know the path which is of interest to you. Sometimes, for gates where the inputs are interchangeable (e.g. *AND* gate), depending upon the performance desired on a specific path, a choice might be made, as to which input pin should it be connected to. As an RTL engineer, you need not worry about this level of connectivity. Synthesis tool would do this for you.

3.3.1.3 Specific Directions of the Transitions

Referring again to Fig. 3.4, when the output *Z* rises due to a transition on *A*, the current flows through transistor *P1*. But, if the output falls due to a transition on *A*, the current flows through transistors *N1* and *N2*. Thus, for a specific *path* (say: *A* → *Z* path) in a specific gate (say: *NAND* gate), the delay depends on the directions of the transitions – because different set of circuit elements are involved in the rising and falling of an output. *A rising* to *Z falling* delay values would be different from *A falling* to *Z rising*. For a *combinational* path, the following combinations may be possible:

- Input rising → Output rising
- Input rising → Output falling
- Input falling → Output rising
- Input falling → Output falling

Not all of these paths are always possible.

A *positive unate* path (such as *AND, OR, buffer* etc.) has only the following 2 types of paths possible:

- Input rising → Output rising
- Input falling → Output falling

A *negative unate* path (such as *NAND, NOR, inverter* etc.) has only the following 2 types of paths possible:

- Input rising → Output falling
- Input falling → Output rising

A *non-unate* path might have all 4 types of paths possible. Consider a MUX, where *D0* is at *0* and *D1* is at *1*. Consider the MUX's *select* to be at *0* – thus selecting *D0*. When the *select* goes to *1*, the output goes *0*→*1*. So, *select* rising → Output rising. Now, if the *select* goes to 0, the output again goes to *0*. So, *select* falling → Output falling.

Now, consider that for the same MUX, the *D0* is at *1* and *D1* is at *0*. You can easily work out that this situation will cause: *select* rising → Output falling and *select* falling → Output rising. So, all 4 types of paths are visible for a MUX's *select* to output path.

Or, only two types of paths might be possible for some non-unate paths. Examples of only 2 possible paths are flop's clock pin to output. The only possible delay paths are:

- Clock rising → Output rising
- Clock rising → Output falling

In the world of *characterization* or STA, such paths along with specific combinations of input and output transition directions are also called *timing-arcs*, or, simply *arcs*. Within a gate, each *arc* would have its own unique delay and output slew specification. After determining the gate and the specific path, the next thing to know is the specific arc for which the delay value needs to be computed.

3.3.1.4 Conditions on Other Pins

Sometimes, the delay through a specific arc is also dependent on the conditions (or, values) on some other pin(s). In such a case, a specific arc might have different delay specifications – each for the conditions of other pins. Examples of such arcs could include:

- Asynchronous clear to flop output when Clk=0
- Asynchronous clear to flop output when Clk=1

If the arc of interest has some such conditions specified, the delay computation has to check whether the specific conditions are met – for the gate under analysis. Only if the specific conditions are met, can the delay/slew from that arc can be used – for the computation. Some engineers consider the condition of other pins to be a part of the arc specification. For example, they would consider the above two delay specifications as two different arcs. While, some characterization engineers consider the above as two different conditions under which an arc has to be specified. An excerpt from a hypothetical library is given below:

```
cell(Flop1) { /* Cell Name – represents a schematic */
....
pin(Q) {
timing() {
....
 rise_transition("template1") { /* Output slew on Q*/
 ....
 }
 cell_rise("template1") { /* Delay on Q */
 ....
 }
```

```
    related_pin : "SET" ; /* Timing for SET to Q arc */
    when : "CLK" ; /*Condition on CLK Pin – under which this timing is valid */
  }
  ...
  }
```

The factors mentioned in this section are discrete. They do not have a continuous range of values. Hence, the characterization is done over all possible values of those factors. For example, each gate will have all its arcs individually characterized.

3.3.2 Continuous Factors

However, the factors mentioned below are continuous. Thus, it is not possible to do the characterization over all possible values. Thus, characterization is usually done for some discrete points – covering the entire range of operation. Delays and output slew are put in the library as a function or table of these discrete points. The STA engine determines the delay/slew at the point of interest through combination of interpolation, extrapolation or *derating*. Derating is explained later in Section 3.3.2.8.

3.3.2.1 Load on the Gate

The delay of a gate depends on the load that the gate's output is seeing. The load seen by the output is a sum of:

- Load of all the inputs that this output has to drive
- Load of the interconnect wires which connect the output to all other portions of the circuit
- If this is a tristated output, then, there might be other tri-state cells' output also connected to this output. Those outputs also put a load.

Usually, with an increasing load, the delay also increases.

3.3.2.2 Input Slew

The delay of a gate depends on the transition time at the input of the gate. The transition time seen at the input of the gate is dependent on:

- Transition time at the previous gate – that is driving this input
- The *interconnect*, which determines how the slew gets modified as the signal traverses over the interconnect wire from the driver gate till the input pin
- For a primary input, the input slew (or, some other factor – from which slew can be directly determinable) is specified. Such factors might include:

- *Drive strength* of the driver (from the drive strength and the load seen by the driver, the transition time can be computed)
- The driver cell itself (from the name of the driver, its drive strength can be picked up from the library)

Usually, with an increase in transition time, the delay also increases. An excerpt from a hypothetical library is given below:

```
cell(Flop1) { /* Cell Name – represents a schematic */
....
pin(Q) {
timing() {
....
 rise_transition("template1") { /* Output slew on Q*/
  index_1 ("0.02, 0.5, 1.0,4.0"); /* Range of output load */
  index_2 ("0.01,0.5,0.9,1.3,2.5"); /* Range of input slew */
 ....
 }
 ...
 }
```

3.3.2.3 Interpolation/Extrapolation

In most libraries, the delays and output slew are specified as a 2-dimensional table. One dimension is the output load, and the other dimension is the input transition time. The table entries are the delay (or, output slew) values themselves. Such table based delay (or, slew) specification is also called as *Piece-Wise-Linear* model, because, the delay (or slew) is considered to be linear between any two consecutive entries of the table. The corresponding excerpt from a hypothetical library is given below:

```
rise_transition("template1") { /* Output slew on Q*/
  index_1 ("0.02, 0.5, 1.0, 4.0"); /* Range of output load */
  index_2 ("0.01, 0.5, 0.9, 1.3, 2.5"); /* Range of input slew */
  values {"0.1, 0.2, 0.3, 0.4, 0.5",\
          "1.1, 1.2, 1.3, 1.4, 1.5",\
          "2.1, 2.2, 2.3, 2.4, 2.5",\
          "3.1, 3.2, 3.3, 3.4, 3.5");
 }
```

If the delay (or, output slew) has to be determined corresponding to load L, and input slew S, the value can be determined through Interpolation. The procedure below explains the way to compute the delay values. The computation for output slew is exactly similar.

Two values on the output-load axis are chosen such that one of them is just lower than *L*, and the other one is just higher than *L*. Call these Load values as: *L1* and *L2*. Similarly, two values on the input-slew axis are chosen such that one of them is just lower than *S*, and the other one is just higher than *S*. Call these slew values as *S1* and *S2*.

Now, from the table determine the delay values corresponding to the following combinations:

- *S1, L1*: Say, delay value is *D11*
- *S1, L2*: Say, delay value is *D12*
- *S2, L1*: Say, delay value is *D21*
- *S2, L2*: Say, delay value is *D22*

Figure 3.5 shows how interpolation is done. D11, D12, D21 and D22 represent the various delay values obtained from the table. L1, L2 and S1, S2 represent the values on the Load and Slew axes that form the smallest bound around the point of interest (L,S).

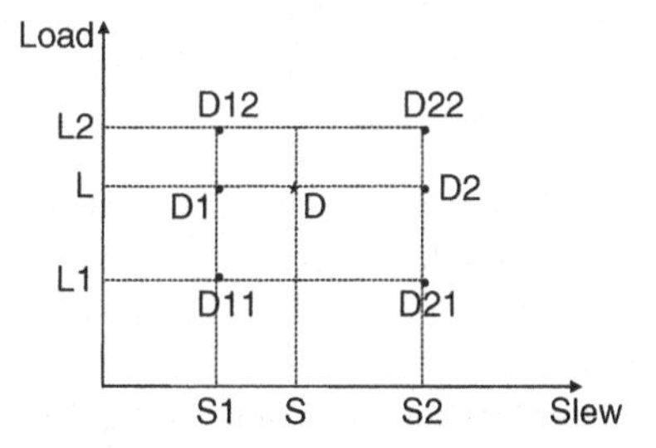

Fig. 3.5 Two dimensional interpolation

Now, the delay of interest can be computed using a series of interpolations. Using coordinates *(S1, L1)* and *(S1, L2)*, *linear interpolation* is used to find the delay corresponding to *(S1, L)*. Say, the value obtained is *D1* – through interpolation of *D11* and *D12*. Similarly, using coordinates *(S2, L1)* and *(S2, L2)*, *linear interpolation* is used to find the delay corresponding to *(S2, L)*. Say, the value obtained is *D2* – through interpolation of *D21* and *D22*. So, the delay values *D1* corresponding to *(S1, L)* and *D2* corresponding to *(S2, L)* are known. Now, using coordinates *(S1, L)* and *(S2, L)*, linear interpolation is used once again to find the delay corresponding to *(S, L)*. This value *D* (represented by an asterix in Fig. 3.5) is the value of interest – obtained through interpolation of *D1* and *D2*. Actually, *delay calculators* are able to determine the values in just one go, rather than going through 3 steps process.

If the value *L* exceeds the largest load value in the delay table, then, interpolation cannot be used. In such cases, the two largest load values and the corresponding delays are used to find the point of interest through *extrapolation*. Similarly, if the value *L* is smaller than the smallest load value in the delay table, then, extrapolation is done, using the smallest two values of load.

Similar to output load, extrapolation might also be used for slew if the input slew falls outside the range over which the delays have been characterized. A need for extrapolation means using the gate in a range – for which it was not even characterized. Usually, this is indicative of using the gate outside the range for which the gate was designed.

3.3.2.4 Temperature

Usually, delay increases very marginally with an increase in *junction temperature*. Junction temperature refers to the temperature within the silicon. In general, the junction temperature is higher than the *ambient temperature* – which refers to the temperature of the environment – in which the chip is being used.

3.3.2.5 Voltage

Usually, delay increases significantly with a decrease in supply voltage. In this context, supply voltage refers to the voltage at the gate. There might be a slight drop in voltage by the time it reaches the gates – compared to what was applied at the supply of the chip, due to a drop through the power rails.

3.3.2.6 Process

The manufacturing process involves some statistical variations. Some portions of the die might receive very strong doping, while, some other portions might receive a weaker doping. These statistical variations are called *process variations* – for lack of a better word. Though, the process variations are continuous, some of the extreme points are given a name. These names could be of the form *SNSP* (Strong N, Strong P), *WNWP* (Weak N, Weak P), *SNWP*, *WNSP*, etc. Or, the names could be of the form *Fast*, *Slow*, *Typical*, or, they could be of the form *Best*, *Worst*, *Typical* etc. Strong denotes smaller delay. Similarly, Fast denotes smaller delay, and, so does Best. On the same lines, Weak, Slow or Worst denote higher delay.

3.3.2.7 Operating Condition

A specific combination of Temperature, Process and Voltage is called *Operating Condition*. In some literature, it is also called as *Timing Corner*. Sometimes, some widely used combinations might be given names, such as *WCMIL* (Worst Case Military), *BCMIL* (Best Case Military), *WCIND* (Worst Case Industrial), *BCIND*, *WCCOM* (Worst Case Commercial), *BCCOM* etc. Some other naming styles might also be used. So, when somebody specifies one of these names, it refers to a combination of Temperature, Process and Voltage.

You might have to do the Timing Analysis at many different operating conditions – to ensure that the device would be able to operate at the desired frequency under all those various combinations of conditions. For devices having applications in a wide range of environment, you might have to cover even up to 20 timing corners.

An example excerpt from a hypothetical timing library shows the operating condition specification:

```
operating_conditions("BCMIL") {
  process     : 0.74 ;
  temperature : -40. ;
  voltage     : 3.465 ; /* 3.3 V technology */
}
```

3.3.2.8 Derating

Characterization (over a range of output load and input slew) is done for various *Operating Conditions*. There would be a separate library for each of these operating conditions. For each of these operating conditions, some more values are chosen – by varying just one parameter slightly, and, delay would be measured again to see the impact of this variation.

For example, consider a library was characterized at 25°C, 1.1 V and SNSP. Now, some more measurements would be made at 20°C, 1.1 V and SNSP (i.e. only the temperature has been changed). The delay differential between the two sets of measurements gives the variation in delay due to 5°C. Using simple arithmetic, we can determine the delay differential per degree Celsius. This is called the *Temperature Derating Factor* for delay. Similarly, the *Voltage Derating Factors* and *Process Derating Derating factors* can be determined for delay, output slew, and, many other parameters of interest.

For computation of delay at a specific operating condition, the library characterized at the nearest operating point is chosen. Say, the delay at that point is determined (through interpolation explained in earlier section) to be *D0* – from this library. Now, the actual delay (*D*) would be computed using derating factors:

$$D = D0 * [1 + k_t (T-t)] * [1+k_v (V-v)] * k_p$$

Where,

k_t = temperature derating factor
k_v = voltage derating factor
k_p = process derating factor
T = Temperature at which delay is to be computed
t = temperature at which delay was characterized
V = Voltage at which delay is to be computed
v = voltage at which delay was characterized

Usually, k_t is a slight positive value, indicating marginal increase in delay per degree Celsius rise in temperature. Comparatively, k_v is a larger negative value, indicating significant decrease in delay per volt rise in supply. Excerpt from a hypothetical timing library shows derating values:

k_temp_rise_propagation : 0.001 ;
k_temp_fall_propagation : 0.001 ;
. . . .
k_volt_rise_propagation : -0.320 ;
k_volt_fall_propagation : -0.320 ;
. . . . .

Derating based computations are not very accurate. Hence, derating is applied only for a very small variation *in operating conditions,* compared to the conditions at which the library was characterized.

Understanding of Section 3.3 *allows you to better understand the need for all the various information that you have to provide to an STA tool – in the form of constraints. Also, it allows you to understand the details of an STA report. And, most importantly, if your design is not meeting the performance parameters that you desire – you know the factors that you can play with, or, what might be causing the bottleneck.*

3.4 Sequential Arcs

Delay relates to an input reaching an output. However, sometimes, there are timing relationships between two inputs themselves, or, multiple edges on the same input. These relationships usually exist only for sequential cells. These are various kinds of requirements that need to be satisfied so that the sequential cell behaves reliably. If any of these requirements fail, the value stored cannot be reliably predicted. Since these are timing requirements that have to be met, these are also referred to as *timing checks*. These checks are specified in tables similar to that of delay or slew, except that instead of input slew and output load, the table entries are based on input slews of the signals – between which the timing check has to be made. These arcs are described in following sub-sections.

3.4.1 Pulse Width

This refers to the minimum width of the pulse on clocks and asynchronous pins, so that the sequential cell can register the clock or an assertion of the asynchronous pin. For a clock pin, this check might refer to both Low Pulse as well as High Pulse. For an asynchronous pin, this refers to only the assertion level Pulse, i.e. for an active Low asynchronous reset – there would be a requirement of minimum width for a Low Pulse. If a pulse is applied – which is of duration shorter than the minimum Pulse Width, the output may not go to the desired level i.e. clock might not be able to capture the new data, or, the asynchronous signal might not get asserted. Figure 3.6 shows an example.

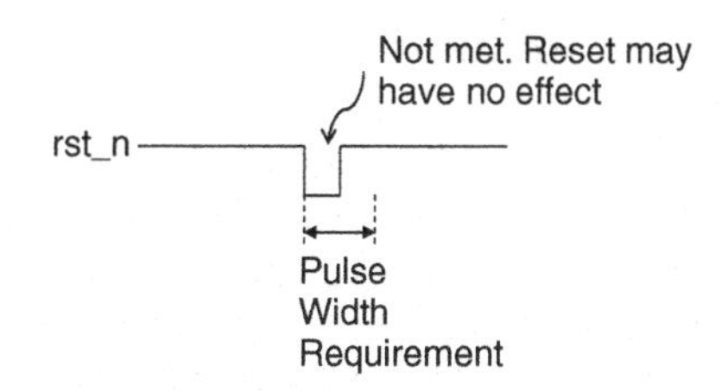

Fig. 3.6 Pulse width requirement

3.4.2 Setup

This refers to the minimum duration *before* the triggering edge of clock, by which the data should have come and stabilized. Alternately, this refers to the time before the clock edge and till the clock edge, during which the data should not change. If the data changes within this duration, there is no guarantee that the new data would get captured in the sequential device. Figure 3.7a shows an example.

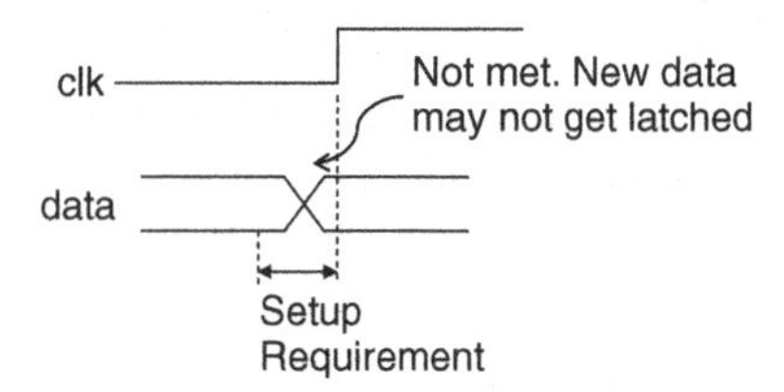

Fig. 3.7a Setup requirement

3.4.3 Hold

This refers to the minimum duration *after* the triggering edge of clock, till which the data should retain its stable value. Alternately, this refers to the time after the clock edge – starting from the clock edge, during which the data should not change. If the data changes within this duration, there is no guarantee that the new data would not interfere with the data being captured within the sequential device. Figure 3.7b shows an example.

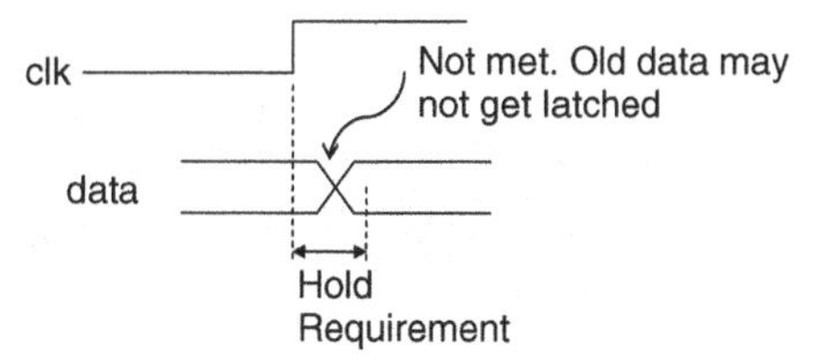

Fig. 3.7b Hold requirement

Together, setup and hold define a window around the clock edge, within which the data should remain stable, and, this stable data would be latched in the sequential device. If there is a change in value during the window defined by setup and hold, there is no guarantee as to what data would get latched.

3.4.4 Recovery

This refers to the minimum duration *after* the de-assertion of an asynchronous control signal, during which the clock's triggering edge should not come. A good way to understand this is: Suppose the asynchronous control signal has been de-asserted. However, the circuit is still not recovered from the effect of the asynchronous control signal. So, if a clock's triggering edge comes – before the circuit could fully recover, the clock edge may not really trigger the circuit. Figure 3.8 shows an example for recovery check.

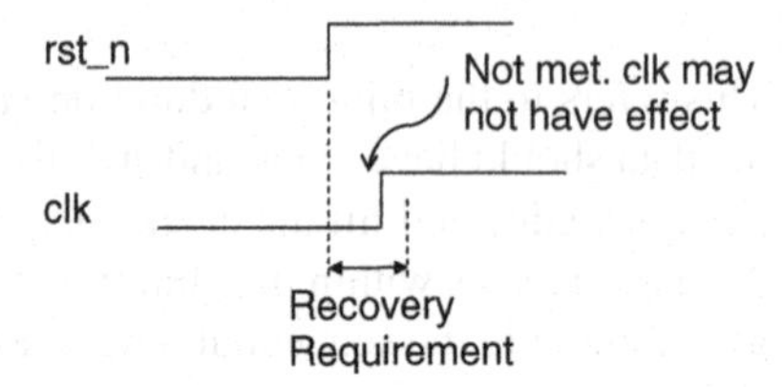

Fig. 3.8 Recovery requirement

Looked alternately, it also means, the time before the clock triggering edge, by when asynchronous control pin should have been de-asserted. Hence, this might be also thought of as a setup check.

3.4.5 Removal

This refers to the minimum time *after* a clock's triggering edge – for which an asynchronous control pin should remain asserted, so that the clock's triggering edge does not have any effect, and the asynchronous control pin decides the state of the sequential element. Figure 3.9 shows an example.

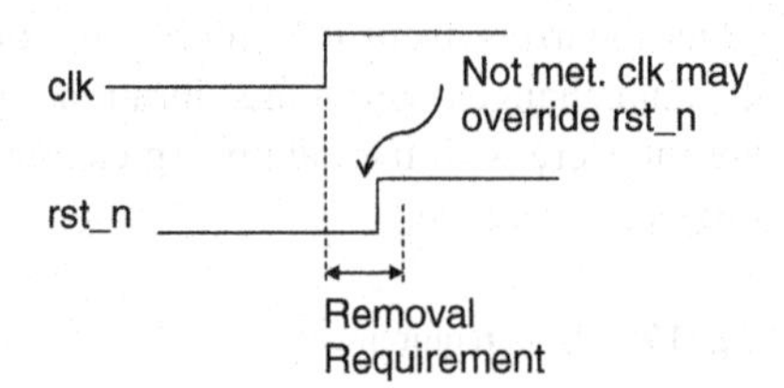

Fig. 3.9 Removal requirement

As can be seen, this might also be thought of as a hold check.

3.5 Understanding Setup and Hold

In this section, setup and hold is explained in more details. Specifically, how/why does setup and hold requirements appear for a sequential device. The same analysis can be carried further to understand how each of the other timing check requirements arises. Consider the circuit shown in Fig. 3.10.

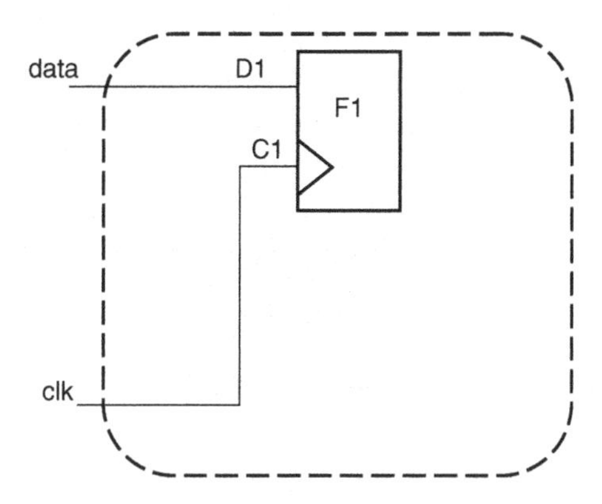

Fig. 3.10 Setup and hold requirement

The boundary represented by the outermost broken line represents a sequential device – for which the setup/hold requirement will be worked out. *F1* represents an ideal flop, i.e. with *0 setup* and *0 hold* requirement. *D1* represents the delay from the input pin *data* till the data inputs of the flop *F1*. *C1* represents the delay from the input pin *clk* till the clock input of the flop *F1*.

3.5.1 Understanding Setup

For the sake of simplicity – first assume that $C1 = 0$. Assume that *data* and *clk* arrive simultaneously. *clk* will reach the flop's clock terminal instantaneously. However, *data* has still not reached the corresponding data terminal. So, if you are looking at the boundary, you see that the value available at *data* at the edge of *clk* is not really getting captured in the device. For you to ensure that the desired value is captured at the flops, you have to ensure that the value is available at *data* – slightly before the triggering edge of clock appearing at *clk*. Actually, value has to be available at *data* at time *D1* before the triggering edge appears at *clk*.

However, in practical situations *C1* itself is also not *0*. So, if you apply the delay of *C1*, the signal is required to be at *data* at time *(D1–C1)*. Figure 3.11 explains the same concept using timing diagrams.

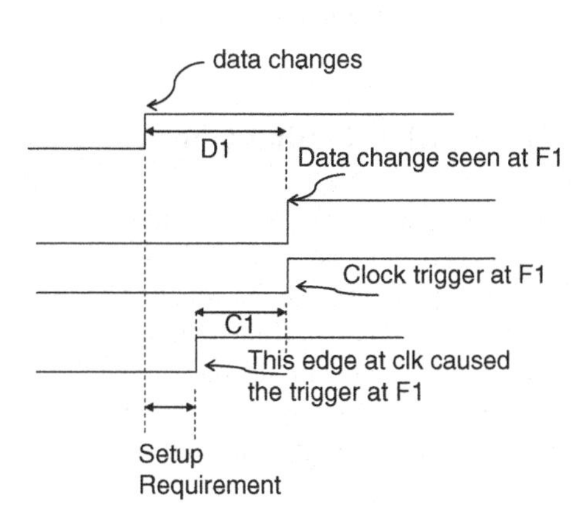

Fig. 3.11 Derivation of setup requirements

Sometimes, there might be multiple data paths or clock paths inside the boundary of the device. In order to come up with the most pessimistic requirements for such devices having multiple data or clock paths, you can say the setup requirement to be:

Setup = Max(all data path delays) – Min(all clock path delays)

3.5.2 Understanding Hold

This time, for the sake of simplicity – assume that *D1 = 0*. Assume that signals arrive at *data* and *clk* simultaneously. New value at data will reach the flop's data terminal instantaneously. However, *clk* has still not reached the corresponding clock terminal. And, by the time, the triggering edge is available at the flop, the new value of *data* is already there. So, if you are looking at the boundary, you see that the value of *data* at the edge of *clk* has been overwritten by a new value that arrived subsequent to the clock edge.

For you to ensure that the desired value is captured at the flops, you have to ensure that the same value continues to remain available at *data* – slightly after the triggering edge of clock appears at *clk*. Actually, value has to be kept at *data* till time *C1* after clock appears at *clk*. However, in practical circuits *D1* itself is also not *0*. So, if you consider the delay of *D1*, the signal is required to remain at *data* till time *(C1-D1)*.

And, if you consider multiple data paths and clock paths inside the boundary of the device, you can say the hold requirement to be:

Hold = Max(all clock path delays) – Min(all data path delays)

The above explains how setup and hold requirements arise for a complex sequential cell. The same principle works within the simplest flop also, because, the way a flop is constructed, there are atleast 4 clock paths – controlling various transmission gates. And, the delay differentials on all these data and clock paths within the flop cause setup and hold requirements.

3.6 Negative Timing Check

If the data path *within* the sequential element is much slower (meaning higher data path delays), compared to the clock path, the hold value can be negative. This should be easily apparent from the equation for hold – shown in Section 3.5.2.

Conceptually also, if the data path (within the device) is too slow, then, even if there is a new value on the data pin at the boundary of the sequential cell, this new value will take too long to reach the actual latching element. And, by that time, the clock would have latched the old value.

Figure 3.12 shows the implication of negative hold check.

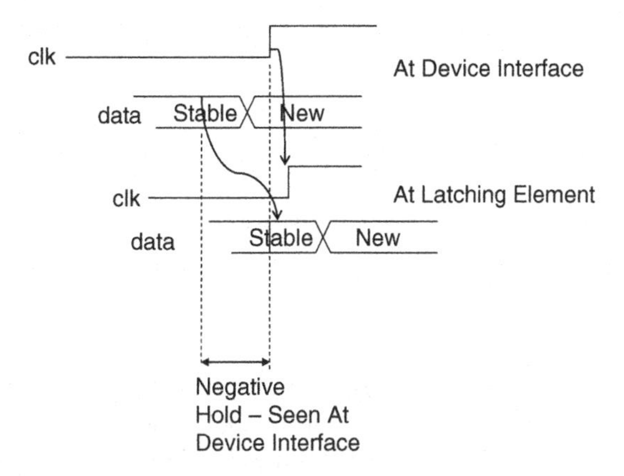

Fig. 3.12 Negative hold

So, even if there is a new value on the *data* pin at the boundary of the sequential device before the triggering edge of the clock, this new value does not interfere with the value being latched. With very small geometries, it is sometimes possible that the hold value in a library can be found to be negative. This negative value of hold means that you can safely put in a new data – even before the clock edge has sampled the previous one.

Similarly, if clock path within the sequential device has higher delay compared to the data path, the setup value can be negative. Negative setup means, the flop can latch a data value which has come after the triggering edge of the clock.

However, since the clock path is usually faster, hence, negative setup is not usually encountered, though, negative hold is rather common. Even though, individually setup or hold can be negative, the sum of setup and hold cannot be negative. If we consider the equations for setup and hold, shown in the previous sections:

Setup + Hold = Max(clock path) + Max(data path) – Min(clock path) – Min(data path)

It is obvious that the value cannot be negative. Even in terms of physical significance, it is difficult to imagine the physical implication of negative (setup + hold).

3.7 Basic Analysis

Consider the circuit shown in Fig. 3.13. This is a generic representation of any circuit. It has combinatorial path from input to output; path from register to register; path from input to register and path from register to output.

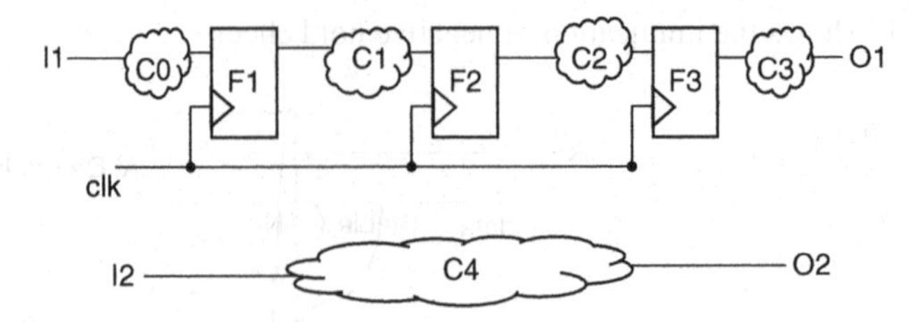

Fig. 3.13 Generic representation of a circuit

Assume:

P = Clock Period
$C0, C1, C2, C3, C4$ etc. = Delay through respective combinatorial clouds
T_I1, T_I2 = Time for the signal to arrive at inputs I1 and I2 respectively
D_Q = Delay for the signal to reach Q, from triggering edge of Clock. This value would be different for different instances of the flops ($F1$, $F2$ and $F3$)
S, H = Setup and hold requirement of the corresponding flop

So, the time at which signal reaches $O2 = T\_I2 + C4$
For $F1$ to capture data reliably:

$T\_I1 + C0 \leq P - S1$ // so that the data is available by the time next clock edge occurs
$T\_I1 + C0 \geq H1$ // so that a new data does not corrupt the sampling of the previous value

This is also expressed as:

$Setup\ slack = P - S1 - T\_I1 - C0$
A positive setup slack value means the setup requirement has been met.
$Hold\ slack = T\_I1 + C0 - H1$
A positive hold slack value means the hold requirement has been met.

For $F2$ to capture data reliably:

$D\_Q1 + C1 \leq P - S2$ and
$D\_Q1 + C1 \geq H2$
Alternately,
$Setup\ slack = P - S2 - D\_Q1 - C1$
$Hold\ slack = D\_Q1 + C1 - H2$

The checks for each flop to see (and ensure) that the setup slack is positive is called *setup analysis*. This uses the maximum delay values for data path, and minimum delay values for the sampling clock path. In the context of this discussion, data path also includes the clock on the launching device. It is worth noting that setup analysis uses clock period also. That means, if a device is failing setup analysis, the device would be able to work at a lower frequency (i.e. higher clock period).

It is possible that within the same chip itself, different portions might have variations in junction temperature, or, supply voltage or manufacturing related parameters. This is called *On Chip Variation.* Thus, the same clock signal might have different delays – on different parts of the chip. Thus, the clock on the launching device needs to be considered as the slowest, while, the clock on the sampling device would be considered the fastest. This differential treatment of clocks on the two devices is done to ensure that the path meets setup – even after considering the *On Chip Variation.*

The check for each flop to see (and ensure) that the hold slack is positive is called *hold analysis.* This uses minimum delay values for data path, and, maximum delay values for the sampling clock path. Hold analysis is not dependent on clock period. Hence, if a device fails hold analysis, changing the clock frequency will be of no help!!! For hold analysis also, the clock delays on launching and sampling flops are considered different – to account for *On Chip Variation.*

Since derating is not very accurate, hence, different derating factors might be applied for the data path and the sampling clock path – to achieve a pessimistic analysis.

The time at which signal reaches *O1 is D_Q3 + C3.* If this signal has to be sampled by another flop driven by the same clock, then, it can have an external delay of *P-D_Q3-C3.* This means that the sum of routing delay (from this output till the input of the next stage), delay through the combinatorial block till the next sampling sequential element and the setup requirement of the sampling element has to be within *P-D_Q3-C3.*

The information related to clock, input signal arrival time, transitions times at the inputs, output signal required time, load at the outputs etc. are all available as user input – specified in the form of *SDC.* The exact format of *SDC* which specifies all this information is beyond the scope of this book. You can get more details on SDC from Synopsys' web-site through their TAP IN programme.

As an RTL designer, you should know the input arrival times. This decides the amount of logic that you can put in *C0* – viz – between the inputs and the first level of flops. Similarly, you should know the output required times. This decides the amount of logic that you can put in *C3* – viz – between the last level of flops and the output. The output required time depends on the routing delay and the delay within the next block – before it gets captured by a register.

You can also decide to register all your inputs as well as all your outputs. This allows your design to be made totally independent of variations in input arrival times and also output required times. Similarly, the input transition time and the output load should be known – because they impact the delay, and, it is your responsibility to ensure that the RTL code that you write should meet the required timing. You also need to know the period of the clock, because that determines the amount of logic that you can put between a pair of flops. The setup/hold values for specific gates are available from the library. The loads and the transitions times at intermediate points within the design are computed by the timing analysis tool.

3.8 Uncertainty

As shown in Fig. 3.14, *uncertainty* refers to the variation in the actual triggering edge of the clock. Uncertainty occurs due to:

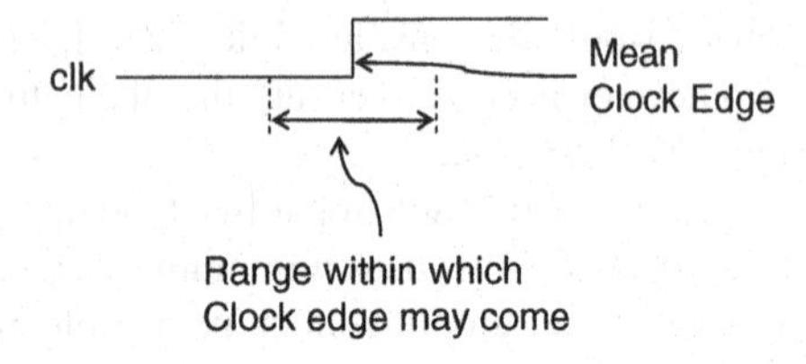

Fig. 3.14 Clock uncertainty

- PLL Jitter
- Skew in different clock paths

Though, the PLL generates clock edges with uniform periodicity, however, due to various physical parameters, there is a slight fluctuation in the time at which these edges occur. This is called *PLL Jitter*. Thus, due to *jitter*, the time between two consecutive active edges of a clock may vary around the mean *Period*. Sometimes, this time might be slightly more than the *Period*, and, at other times, it might be slightly less than the *Period*. The shortest time between two consecutive active edges of a clock can be *(Period – Jitter)*. Similarly, the longest time between two consecutive edges of a clock can be *(Period + Jitter)*.

As a clock has to drive many flops, thus for a given clock edge at the source, all the flops receive the edge on their terminals at slightly different times. The difference in time at which various clock terminals receive the clock edge is called the *Skew*. This *skew* thus defines the differential delay on the clock paths for any pair of flops. Before layout, it is known that all the flops will receive the clocks within a given range (defined by *maximum allowed skew*). However, it is not known as to which flop will receive the clock at exactly what time within the range. Thus, the maximum allowed skew contributes to *uncertainty*.

Suppose, a launch flop gets is clock edge earlier than the capture flop. In that case, the data gets a head start in getting launched. Similarly, if the launch flop gets its clock edge later than the capture flop – the data has already lost some time before it can get launched. Thus, because of *skew*, there can be a slight increase or decrease in time available for the data to be transmitted.

Consider a setup analysis. In the worst case scenario, the time between consecutive active edges of clock would be reduced by *jitter*. Also, the time available for data transmission could get reduced by allowed *skew*. This will cause setup slack to be reduced by an amount equal to the maximum range of variation (i.e.

Jitter + Skew) in clock edge. You should specify the combined impact of both *jitter* and *skew* as *uncertainty*.

For hold check, the same edge of clock is used for both launch as well as capture. Thus, PLL *jitter* will not cause any differential in clock edges – for the hold check, because, both the launching edge and the capturing edge will move equally in the same direction. The *uncertainty* is only due to *skew*. So, for hold analysis, you should specify only the allowed *skew* value as *uncertainty*. For both setup and hold analysis, the slack will be reduced by an amount equal to the specified *uncertainty*; though, you may decide to use smaller value for uncertainty during hold analysis compared to the value used for setup analysis.

Before layout, the actual *skew* value between the launching and capturing flops is not known. Hence, maximum possible *skew* is used for *uncertainty*. However, after layout, the actual *skew* value is already known. Hence, the delay differential no longer contributes to *uncertainty*. Whatever *skew* is there – can be computed, and, actual *skew* value can be used rather than the maximum possible value. Thus, during post-layout timing analysis, you should reduce the *uncertainty* values – corresponding to the skew values used during pre-layout timing analysis.

If you try to do a post-layout timing analysis using the same value of *uncertainty* (as used during pre-layout), there is a strong possibility that some slacks will come out to be negative (i.e. timing violation). This is because, the actual *skew* is already being considered. And, beyond that, the *uncertainty* will further reduce the slack. For critical paths, which might be just about meeting the timing, this additional reduction might cause the timing to be not met.

The *uncertainty* information is specified by the user through *SDC* commands. *SDC* commands allow for *uncertainty* to be specified for:

- each clock
- pair of clocks
- uniquely for setup or hold
- combinations of above

There are separate SDC specifications for *pre-layout* and *post-layout*. This allows a user to specify different uncertainty values for *pre-layout* and *post-layout*.

3.9 STA Contrasted with Simulation

Towards the beginning of this chapter, we have seen the difference between STA and dynamic simulation mostly in the context of exhaustiveness of timing check, as well as basic approach – in terms of need to apply vectors. Besides these differences, there is a fundamental difference in how STA treats setup and hold, compared to how simulation treats setup and hold.

In the context of simulation, setup and hold define a window, within which the signals should not change. On the other hand, in the context of STA, setup defines a value – before which all transitions should have arrived, and, hold defines a value – before which none of the transitions should have arrived. Because of this fundamental difference in definitions of setup and hold, it is possible to get anomalous results. Consider the circuit shown in Fig. 3.15.

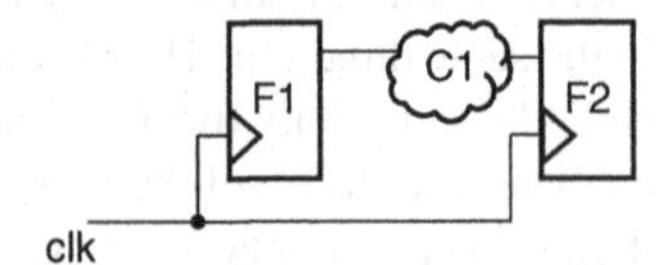

Fig. 3.15 Circuit for understanding simulation vs. STA

Let the clock period be *10*. Let the setup and hold requirements of the flop *F2* be *1* each.

3.9.1 Setup Violation in STA, No Violation in Simulation

Consider a situation, where the delay from the launch edge of the clock till *F2* (i.e. F1's Clk → F1's Q → F2's D) is *13*. So, the change at *F2* would be visible at *13* – which is clearly outside the constraints of setup and hold window. Hence, simulation will not give any violation. However STA makes the check differently. It checks whether the signal will reach before the setup time for *F2*. So, since the signal does not reach *F2* within *9 (it's reaching at 13 i.e, 4 s later),* hence, STA will report a setup violation. Actually it reports a negative slack for setup check.

3.9.2 Setup Violation in STA, Hold Violation in Simulation

Consider a situation, where the delay from the launch edge of the clock till *F2* is *10.5*. So, the change at *F2* would be visible at *10.5* – which is clearly within the *hold* window. Hence, simulation will give a *hold* violation. However STA makes the check differently. It checks whether the signal will reach before the setup time for *F2*. So, since the signal does not reach *F2* within *9,* hence, STA will report a *setup* violation. Actually it reports a negative slack for setup check.

3.9.3 Hold Violation in STA, Setup Violation in Simulation

Consider a situation, where the clocks on the launching flop and the capturing flop are skewed – in a manner that the launching flop receives its clock earlier than the capturing clock. The delay from the launch edge of the clock till *F2* is very less – lesser than the clock skew. Figure 3.16 shows the corresponding timing diagram.

So, the change at *F2* would be visible just before the clock edge – which is clearly within the *setup* window. Hence, simulation will give a *setup* violation. However STA makes the check differently. It checks whether the signal will reach after the

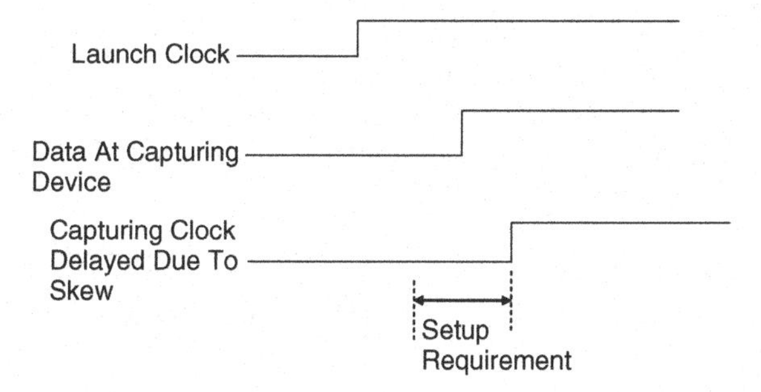

Fig. 3.16 Setup in simulation, hold in STA

hold time for *F2*. So, since the signal reaches *F2* before the hold time, hence, STA will report a *hold* violation. Actually it reports a negative slack for hold check. This is a situation of *feedthrough*. (Feedthrough has two meanings. In this context, feedthrough has the meaning as explained in Section 2.3).

If we extend this further, so that the clock skew is much higher compared to the data path delay, then, the change might reach the capturing flop much before the clock edge – so much so that – it falls even before the setup window of the capturing flop. In such a case, simulation wont report any timing violation. However, STA would report the hold violation.

Usually, clock skews are very small. Hence, it is quite unlikely that a data path delay is less than the clock skew. Thus, this kind of situation is not very likely. The only place where such situations might occur are for shift-register kind of circuits, where, the data path delay is also very less. Thus, for shift registers, special attention is given to reduce the clock skew to bare minimum. As an RTL designer, you should let your layout engineer know about any shift register that you have in your design – so that the clock skew can be paid extra attention for the shift registers. Else, there would be a risk of *feedthrough*.

3.10 Accurate Timing Simulation

The discussions in the previous section imply that the simulation is able to accurately consider the delays. However, simulation models do not model delays as functions of input transition or output load. Hence, the delay values in simulation models are not very accurate.

For accurate timing to be considered during simulation, the delay values are determined through timing analysis. Timing analysis tool computes accurate delay values for each instance of the gate – considering input transition, load, temperature, voltage, process etc. The tool also computes the interconnect delay. All these accurate values are put into a file, in a format called *Standard Delay Format* (*SDF*). During gate level simulation, the *SDF* file is also fed to the simulator. The simulators are able to pick up accurate delay and timing check values from the *SDF* file, and, use those during the simulation. The process of simulators picking up the delay values from the *SDF* is called *SDF back annotation*, or, simply *back annotation*. Some people also refer to this as *timing annotation*. A gate level simulation that involves accurate timing is called *Full Timing Gate Level Simulation (FTGS)*. Figure 3.17 shows the flow for *FTGS*.

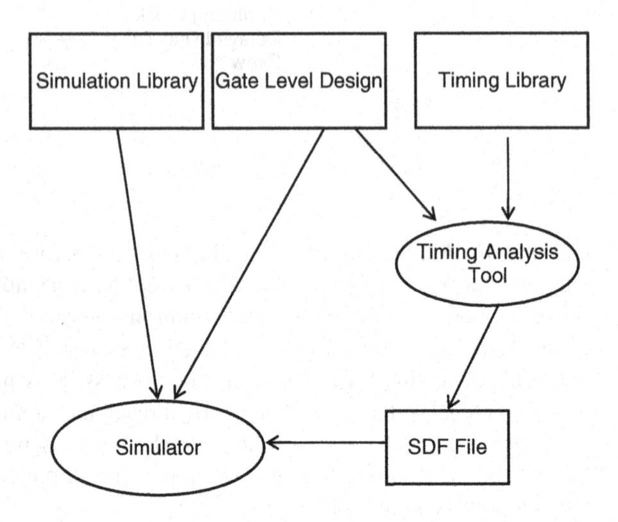

Fig. 3.17 FTGS flow

3.11 Limitations of Static Timing Analysis

In Section 3.1.1, we have already seen limitations of dynamic timing analysis – based on simulation. Section 3.9 also gives several examples that indicate how a simulation based timing analysis could sometimes be misleading. All discussions of STA explain how this is an exhaustive check, as compared to a simulation based timing check. However, still many designs fail timing – even though, they were STA clean.

Actually, the concept of STA being exhaustive check provides a sense of security. However, it has to be realized that STA is driven largely by SDC commands or constraints, which specify the external environment of the design. Also, STA depends a lot on many user specifications, such as clock waveforms, a specific net or pin being at a fixed value etc.

If these specifications are different from what the design is actually going to encounter, there is a strong likelihood that the actual manufactured device might not meet the timing even though the STA is clean. These specifications expressed through SDC span across thousands of lines many times. Unfortunately, while there is a lot of importance given to verify the functionality of the design, much lesser importance is given to validate the SDC. A mistake in SDC might not get caught during STA – giving a false sense of security. Hence, it is of utmost importance that SDC also should be good – so that the results of STA are really reliable.

Some sophisticated design methodologies use FTGS to validate the timing constraints used for STA to minimize this risk. Other design methodologies have started depending upon specialized checkers – to ensure completeness and correctness of these constraints.

3.12 SSTA

SSTA stands for *Statistical Static Timing Analysis*. This is a relatively new concept. It differs mainly from STA in the sense that STA gives a Pass or Fail decision, as to whether or not a chip will meet the timing. While doing this analysis, STA considers the most pessimistic analysis – at the given timing corner. However, SSTA instead of giving a simple Pass or Fail provides statistical numbers – in terms of how many of the devices might meet the timing, and, how many of the devices might miss the timing by some given value. So, if the same device can also be used for some other slower (and, lower cost) applications, you might still send in the design for fabrication. However, SSTA is too complex. And, as of now, it has not gained too much popularity.

3.13 Conclusion

A good understanding of factors that impact timing allows you to write an RTL, for which timing is relatively easier to meet. Besides, you would be able to identify portions of design that might require special attention to timing during layout. You might be able to communicate those special care abouts to the layout engineer.

One of the most important benefits of having a good understanding of timing is that it will help you ensure that your SDC (which is used for driving STA tools, synthesis tools and P&R tools) is clean. Even if you are not directly responsible for any of the above mentioned activities, the fact is that SDC accuracy and its consistency with RTL is such an important requirement of the whole design process, that, you will have to remain involved in reviewing the SDC and authorizing the changes and edits to the SDCs. You can do an effective job – only if you have a very good understanding of timing concepts.

Chapter 4
Clock Domain Crossing (CDC)

Today's chips have many clocks. Different parts of the circuit operate on different clocks. It is also common to have situations, where, the same portion of the circuit might perform multiple operations – sometimes on different clocks. Hence, it is very common to have data generated from one clock being consumed by some other clock.

4.1 Clock Domain

In different contexts *Clock Domain* might have different meanings. In the context of this chapter, two clocks are considered to be in two different domains, if they are asynchronous to each other. If two clocks are asynchronous to each other, then, the time gap between the edges of these two clocks keeps changing continuously. Consider two clocks, clk1 and clk2, with periods *13* and *10* respectively, as shown in Fig. 4.1.

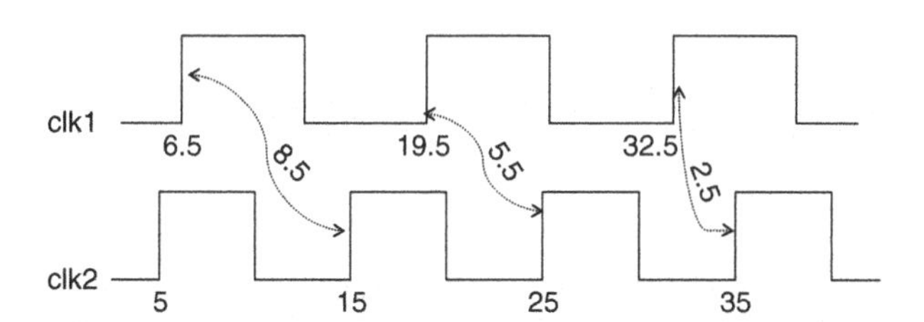

Fig. 4.1 Asynchronous clocks

The time between the edges of *clk1* (leading) and *clk2* (immediately following edge) keeps changing. Hence, *clk1* and *clk2* are asynchronous to each other. When data generated by a clock is captured by another clock which is asynchronous to the clock which generated the data, it is called *Clock Domain Crossing*.

S. Churiwala, S. Garg, *Principles of VLSI RTL Design*,
DOI 10.1007/978-1-4419-9296-3_4, © Springer Science+Business Media, LLC 2011

4.2 Metastability Due to CDC

One of the biggest problems due to CDC is *metastability*. In the previous chapter, we have seen that for a reliable transfer of data between any pair of registers the setup and hold requirements have to be met. However, imagine what happens if the setup and hold requirements are not met. The data capture is not reliable. Not reliable means, it might capture the intended value, or, it might capture the unintended value. Worse, it might not capture any value – and might go *metastable*.

In case of CDC, as seen in Fig. 4.1, the time available for a signal keeps changing for each edge pair. Hence, even if the timing is met for some specific pairs of clock edges, there is quite a likelihood that for some other pair of edges, the setup or hold might not be met. Also, the extent of violation would keep varying across different pairs. Thus, sometimes the setup and hold requirements would be met with sufficient slack, sometimes they would be just met, sometimes they would be just violated and sometimes, there would be gross violations.

4.2.1 Understanding Metastability

Consider the waveforms shown in Fig. 4.2

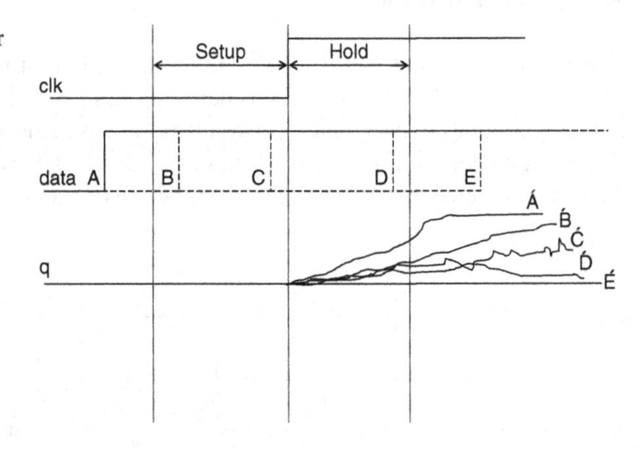

Fig. 4.2 Output response for violating/meeting setup and hold

A, B, C, D and *E* show different points when the data input might change. Á, B́, Ć, D́ and É show the corresponding waveforms at the output of the flop. Transitions at *A* or at *E* are meeting setup and hold requirements. Hence, the output waveform at the flop is also very clean (i.e. reliable). *B* just violates the setup requirement, while, *D* just violates the hold requirement. Hence, B́ and D́ reach the intended value finally, but, they take longer to reach the final intended value. In the previous chapter, we had also seen that the *clk-to-q* delay plays a role in the setup and hold slack for the next flop. So, a higher delay on this flop could cause the setup slack for the next flop to be negative. So, here, the data capture is not reliable, because, though, this flop did eventually reach the intended value – but, it might not be good enough for the next stage. However, the transition at *C* is of interest from metastability perspective. Here, *Ć* does not seem to go anywhere. It will eventually settle somewhere, but, there is no guarantee – to what value, and, more importantly – when.

Consider a hill with steep slopes. If a ball is dropped – such that it falls on the incline, then, it is clear that the ball will roll down the incline towards the foot of the hill – on the same side. Now, consider that the ball is dropped from slightly above the hill. If the ball is dropped significantly towards one side from the top – it would be easy to see that the ball will fall on the same side of the incline and will roll down the same side. This represents the transitions at *A* and *E*, causing the resultant *Á* and *É* waveforms. Consider the same ball being dropped, this time – vertically above the top of the hill – but, moved just slightly to one side. There is a very small likelihood that the ball might cross over the top and reach the other side (say: due to wind or, some other physical disturbance) roll down that incline. Or, it might just roll down the incline over which it was dropped – which is more likely. In either case, it would come to the foot of the hill – rather quickly. This represents the transitions at *B* and *D*, causing the resultant B́ and D́ waveforms. Now, consider the ball is placed very carefully at the top of the hill. If it is placed very carefully, it might just stay there. However, for how long? That depends on how sharp is the peak of the hill. It also depends on any disturbance. It could be wind or, vibrations due to foot-falls of grazing cattles – anything. The ball could theoretically lie there forever – if there was nothing to disturb it. However, some slight external disturbance could cause the ball to roll slightly to one side, so that it reaches the incline – and then, down it rolls – to the foot of the hill.

This is what is metastable – represented by *Ć*. The flop output has got stuck somewhere in the middle. Finally, it will reach somewhere (either a *0* or a *1*) – due to some external electrical disturbance within the device. But, there is no way to know, when will that disturbance occur and which side will *Ć* finally reach. In the current context, *external* means external to this flop.

4.2.2 Problems Due to Metastability

In Chapter 5, we will see that a short circuit path is established between supply and ground, if the input to a set of CMOS transistors is somewhere mid-way between 0 and 1. So, if a flop's output has gone to a metastable state , then, that means a short circuit path gets established across all the transistors that is driven by this specific flop. That means a higher amount of current flow, which means higher power dissipation. However, a much bigger problem is that the same flop output might be interpreted differently – by different parts of the design. Consider the circuit shown in Fig. 4.3.

Say, all the three flops are driven by the same clock with a period of *10*. Say, combinatorial circuit *C1* has a delay of *5*, while, C2 has a delay of *7*. Assume that there is a possibility of *F1's* output going metastable because the source of data for *F1* (not shown in the diagram) is coming from an asynchronous clock. *F1* finally settles

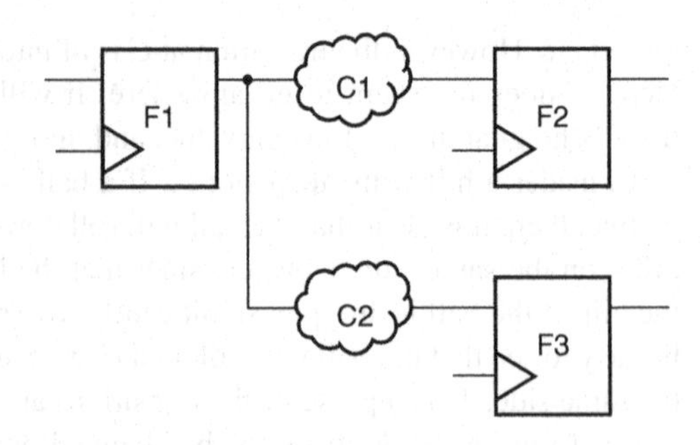

Fig. 4.3 Impact of metastability

down to some value (say) at time *4*. Now, this settled value will reach *F2* at time *9* and hence, *F2* will see the settled value. But, for *F3*, the settled value will reach at time *11*. Hence, the clock at time *10* will get a different value. So, effectively, both *F2* and *F3* which should have seen the same value at *F1's* outputs are actually seeing two different values. The design is obviously expecting both *F2* and *F3* to be seeing the same value for *F1*. Hence, the design might malfunction – thus, giving rise to reliability concerns. Further, depending upon the time at which *F1's* output finally settles and the combinatorial delays *C1*, *C2* etc. there could be setup/hold violations on flops *F2* and *F3* – thereby causing the metastability to propagate further.

4.3 Synchronizer

Synchronizers are used to solve the above problems.

4.3.1 Double Flop Synchronizer

Figure 4.4 shows one of the simplest synchronizers.

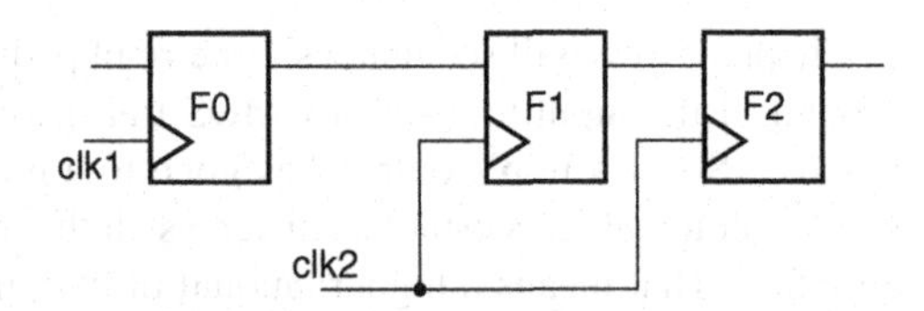

Fig. 4.4 Synchronizer

There is a CDC on flop *F1*. Hence, *F1's* output could go metastable. This output is fed directly to another flop *F2* – which is driven by the same clock as *F1*. *F1* is also called *destination flop*. The second flop *F2* is called the *synchronizing flop*. Now, the output of this second flop can be used for normal device operations. This technique of synchronization is called the *Double Flop Synchronization Scheme*.

Now, even if *F1's* output is somewhere near the middle of *0* and *1*, the issue related to short circuit power is impacting just a few transistors within *F2*. Even within *F2*, depending upon exactly how is the flop designed, the data could be driving *transmission gates*, rather than the *gates* of MOS devices. So, the problem related to excessive power/heat gets controlled. The data at *F1* is sampled by just one device *F2*. So, there is no chance of the same data being read as two different values – because, there is only one device (the synchronizing flop) which is reading this data. Hence, reliability issues are taken care of.

The fundamental concept behind such a scheme is that by the time of the next clock edge, the output of *F1* should hopefully come to a stable state. And, *F2* will be able to capture this stable state, which would now be visible to the rest of the design. There is usually no logic between *F1* and *F2*. This ensures that the delay between *F1* and *F2* is very low. This allows that even if *F1* settles close to the next clock edge, *F2* gets the settled value. You should not put any logic between *F1* and *F2*. By putting any such logic, you effectively reduce the time given to *F1* to settle. And also, the gates forming this logic could see high short circuit, during the duration that *F1's* output is not settled.

It is worth understanding that when you are dealing with metastability due to CDC, it is not important whether the final settled value is at the old value on *F1's* data, or, the new value on *F1's* data. Separate mechanisms anyways need to be put in place to ensure that there is no data loss. Synchronizers are just to ensure that there is no metastability. They are not dealing with data-loss. However, there is still a risk that *F1's* output still does not go to a stable value, even till the next edge of the clock. In that case, *F2* does not see a stable value – and hence, even it might go metastable in the next clock edge.

4.3.2 Mean Time Between Failures (MTBF)

Reliability factors are characterized by *MTBF*. *MTBF* means on an average, whats the time gap – between two failures. So, higher *MTBF* means lesser number of failures within a given time duration, which means more reliable. Adding the additional flop does not remove the chance of failure. It only reduces the chances of failure – which means increases *MTBF*. If there is an application which needs a much higher *MTBF*, you will need to put additional flops beyond *F2* also. Thus, the more flops you add in the series, the more time you are providing to the signal to become stable, and, thereby increasing the *MTBF*. However, you are paying the penalty in the form of latency – your results get delayed by those many cycles.

So, depending upon the *MTBF* desired, a designer might have to compromise between latency and *MTBF*. Sometimes, for relatively low-reliability requirements, you might use just half a cycle for synchronization. This can be achieved by driving the *synchronizing flop* on the negative edge of the clock (assuming, the *destination flop* is triggered on the positive edge of the clock).

4.3.3 Meta Hardened Flops

Sometimes, your cell library might provide for special flops – that have a tendency to come out of metastable state much faster. Such flops are called *meta-hardened flops*. If your library provides for such a flop, you can use such meta-hardened flops for *F1* – which is where the CDC has occured. Use of meta-hardened flops increases the chance that by the end of one cycle, the output has stabilized. This reduces the chances that *F2* will also become metastable. So, using *meta-hardened flops* increase the *MTBF* significantly.

4.4 Bus Synchronization

In the context of this section, *bus* means a set of signals, where the significance of the signals can be interpreted only in the context of the complete set of signals. Just a single bit among the set by itself does not signify anything. Examples could include:

- Vectored Data Lines
- Address Lines
- Control Bus (meaning a set of control signals)

4.4.1 Challenge with Bus Synchronization

In Section 4.3, we saw a simple method to synchronize signals that cross clock domain. However, imagine if a bus is crossing a clock domain boundary. One way is obviously to extend the concept mentioned in Section 4.3, and, thus synchronize each signal individually. In such a case, it is possible that some bits of the bus exhibit their old values, while, some other bits exhibit their new values. This might happen because of either or all of the following conditions:

- Difference in routing delays for different bits could cause these signals to reach at different points with respect to setup/hold window.
- Different bits are being captured by different flops, and, each flop might behave in its own manner – because of metastability. And, as a result, some flop might reach stable state much earlier, and some flop might take much longer to come to a stable state. Also, even the stable state reached could be different for the different flops.

Thus, effectively, when looked in the context of the whole bus – even though the individual bits are synchronized, the data on the bus after synchronization could be totally useless – in the sense that it neither represents the old data, nor does it represent the new data. Some bits are old data, while, some bits are new data – thus, making the final bus value to be garbage.

4.4.2 Grey Encoding

The simplest solution to the above problem is that you could grey encode the bus signals. Grey encoded signals mean that only a single bit changes. Thus, for all other bits, whether the individual synchronizer captures the new data or the old data, it does not really matter, because, either way, it's the same value. Only for one bit (that changed), the new data or the old value is different. However, irrespective of which data is captured for this one bit, the final value of the entire bus is still meaningful. Even in the context of the whole bus, it is not garbage. The value could be the old value or the new value – but, a meaningful value nevertheless.

However, Grey encoding can be used only if the consecutive values have to follow a pattern. If the consecutive values do not follow any pattern, but, are random data, then, grey encoding is also not possible. An example application where such grey encoding might work well is that of a counter. Similarly, an example where such grey encoding would not work is that of a memory read operation. The data being read in two different cycles are from two different locations of the memory, and, there could be no way of ensuring only a single bit change between these two data.

4.4.3 Enable Synchronization Method

A simple solution to this is by using an *enable* signal and that *enable* signal is synchronized – just by itself. The process of synchronization – in effect allows more time for the *enable* signal to settle. Within this extended duration allowed through the process of *synchronization*, it is expected that the individual data bits would have also settled. So, the entire bus can now be captured into the destination flops. Figure 4.5 shows one such circuit based on *Enable Synchronization Scheme*.

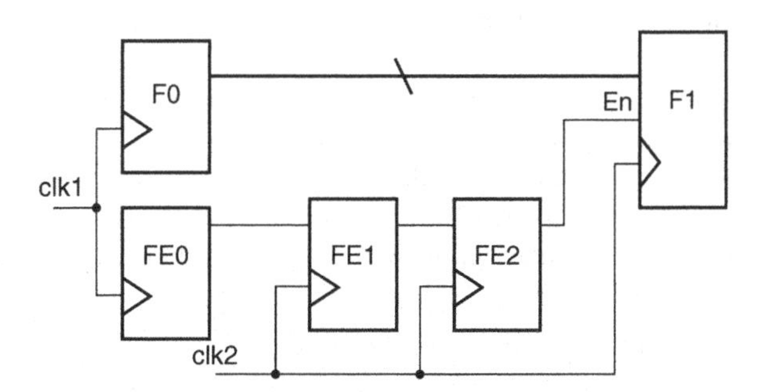

Fig. 4.5 Enabled synchronization

F0 represents a *bank of flops* which generates the bus that will cross the clock domain boundary. *FE0* is another flop driven by the same clock – and it generates the *enable* signal – at the same time as the data is being generated by *F0*. This *enable* signal is synchronized through *FE1* and *FE2*. Till the Enable signal does not reach the register bank *F1* – the arrival of the data inputs on *F1* does not matter. All clocks on *F1* will be ignored – during this duration, due to lack of a valid *enable*. Hence, even if there is a setup/hold violation it does not matter.

By the time, *enable* passes through *FE1* and *FE2*, it is expected that the data bus at the input of *F1* register bank would have settled. At the next edge of *clk2*, the bank *F1* will sample its data – which is all settled to its new value. So, *F1* will sample the new value of the bus. Figure 4.6 shows a slight modification of the above circuit to realize a similar impact.

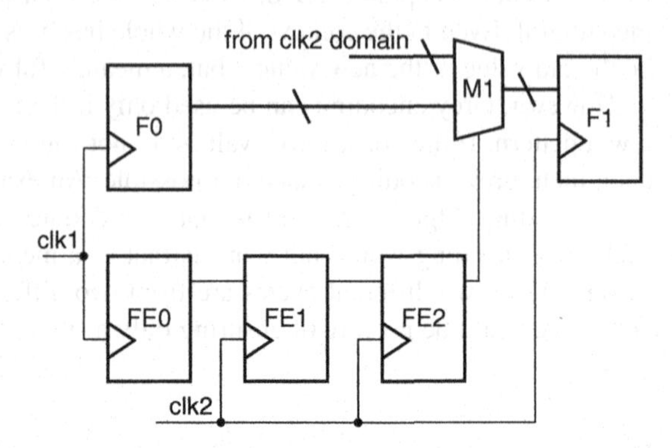

Fig. 4.6 Another scheme for enabled synchronization

The multiplexor *M1* will send data from *clk2* domain into *F1* bank. Hence, this is not a *CDC*. However, when *F0* bank has to send some bussed data to the *F1* bank, an *enable* is also generated (by *FE0*). This *enable* goes through synchronization via flops *FE1* and *FE2*. The process of *enable synchronization* gives enough time for the *F0's* outputs to be settled at the input of *M1*. And, the *enable* signal acts as the *select* of the *multiplexor* in order to pass on the values coming from the *F0* bank. And, at the next triggering edge of *clk2*, the *F1* bank will capture the settled values. It is possible to obtain some more variants of these two schemes. However, the fundamental philosophy is the same, viz:

- Generate a *control* (or, *enable*) signal also along with the bussed signals
- Synchronize the control signal
- The time that is needed to synchronize the *control* signal would usually be sufficient for each bit of the bus also to have settled to its new value
- Use this *synchronized enable* signal to sample the values of the bus – which are settled by now

4.4.4 Cost of Enable Synchronization

For using an *enable synchronization* based method, an additional enable signal has to be generated. This would mean additional circuitry for generating this enable signal. This in turn would mean additional area and power. Further, similar costs also get added for synchronizing this additional enable signal. However, this cost is more than offset, because individual bits of the bus are no longer required to be synchronized individually. So, we can save on one synchronizing flop per bit. And, since we are talking about vectored signals, so, there has to be multiple bits. That means, the savings are in terms of multiple flops.

However, once the bussed signal is generated, it takes 3 triggering edges of the destination clock for the values to be sampled.

- On the first edge, the enable signal goes past FE1
- On the second edge, the enable signal goes past FE2
- And on the third edge, the bussed signal actually gets captured on the destination side register bank.

Contrast this with the double flop synchronizing scheme mentioned in Section 4.3.1. There, the data reaches the destination flop at the second triggering edge of the destination flop. So, Enable Synchronization scheme has an additional latency. But, that's the price that you have to pay – in order to avoid the risk of capturing garbage data.

4.5 Data Loss

So far, you have seen synchronization, which is about mitigating reliability risks due to metastability. However, another risk associated with CDC is related to *Data Loss*. Since the data is being generated by one clock, and, is being consumed by another clock – which is running at a totally different frequency, it is possible that the launching clock might launch a new data even before the capturing flop was triggered to capture the previous data. If such a thing happens, then, the previous data could be lost. Hence, in a design which has CDC, you have to take sufficient precaution against loss of data.

4.5.1 Slow to Fast Crossing

Consider a situation, where the source flop is generating the data at a lower frequency. And, the destination flop is getting triggered by a faster clock. In this case, before the source flop generates another data, the destination flop would have sampled the previous data. Thus, for slow to fast crossing, there might not be a risk of data loss. However, if the destination clock is only marginally faster than the source clock, the data loss risk would still be there. This happens because once the edges of the two clocks are almost aligned they will come very close together for next several cycles. Figure 4.7 shows such a scenario.

The data produced by the first edge of the source clk could not be captured reliably, because of setup violation at the destination clk. The next edge of the

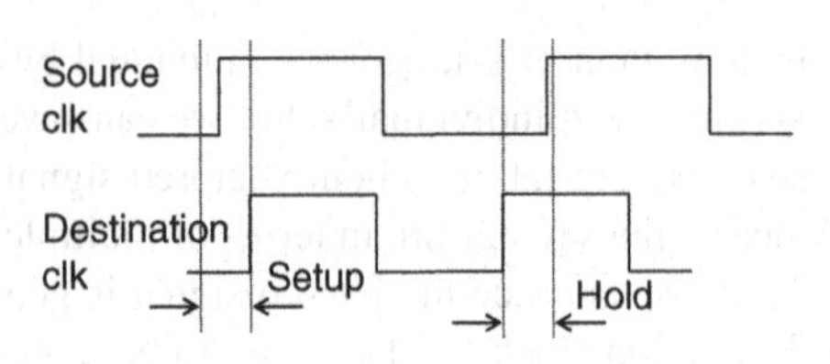

Fig. 4.7 Destination clock faster than source clock

destination clk should capture this data, but, the source flop sends another data – immediately after the second edge of the destination clk, which causes hold violation on the destination flop. Hence, the data launched by the source clk might not be captured reliably at either of the two edges – resulting in loss of data.

Thus, to ensure no loss of data, the destination clk time period should be *sufficiently* less than the time period of the launch clk. *Sufficiently* is required to take care of all kinds of uncertainty. The factors that cause this uncertainty are:

- Setup requirement of the destination flop
- Hold requirement of the destination flop
- Clock Jitter for both launch and capture clock
- Path delay differential for fastest and slowest path from the launch to the capture

So, if the destination clk is sufficiently faster than the source clk (with the above factors already taken into account), there is no risk of loss of data. Section 4.6 explains how to avoid data-losses, where, the clocks involved in the CDC have frequencies that are relatively close to each other.

4.5.2 Fast to Slow Crossing

Consider a situation, where the source flop is generating the data at a higher frequency. And, the destination flop is getting triggered by a slower clock. In this case, before the destination flop captures a data, the source clock would have launched the next data. Thus, for fast to slow crossing, there is always a risk that only intermediate data might get captured, and, several data might get lost. In order to prevent this data loss, it is important to turn off data generation till the capture clock has been able to sample the data.

For crossing between synchronous clocks, where the generating clock frequency is an integer multiple of the sampling frequency, the data loss prevention technique is explained in Chapter 7. For fast to slow crossing between asynchronous clocks, you have to establish a *handshake* based mechanism or protocol between the launch flop and the destination flop. The handshake mechanism ensures that till the first data has been captured reliably, the next data would not be launched by the source flop. Thus, by preventing the launch of the next data, handshake ensures that the first data remains available – till it has been captured. And, only after a data has been captured, the next data would be launched.

Thus, handshake removes the risk of data-loss. However, you pay the cost in terms of additional circuitry for the handshake protocol. Also, there are additional cycles that are consumed in the exchange of Request, Acknowledge etc. which increase the latency.

4.6 Preventing Data Loss Through FIFO

Data Loss can also be prevented *using FIFO (First In First Out)* based mechanism. *FIFO* based mechanism is very useful for either of the two situations:

- The two clocks have very close frequency.
- The data launch is in bursts, i.e. after launching several data in consecutive cycles, it then becomes quiet for several cycles. In such cases, even if the launch clock is faster than the capture clock, the *FIFO* based mechanism can be found to be useful. The data launched in the bursts keep getting stored in the *FIFO*. While, the launch side becomes quite, the capture side keeps picking up the data stored in the *FIFO*.

FIFO is helpful in such cases, because, it acts as an *elastic buffer*. So, for pairs of clocks which are very close in frequency, the launch clock can keep launching data – which gets stored in the FIFO, and, at each capture clock, the next data from the FIFO can be captured.

If the launch frequency is slightly higher than the capture frequency, then, after several cycles, the FIFO might become *FULL*. At this stage, the launch has to be stopped for one or two cycles. Within these one or two cycles (of stopped launch), the capture clock will read a few more locations of the FIFO – making additional space in the FIFO – to resume the launch.

If the capture frequency is slightly higher than the launch frequency, then, after several cycles, the FIFO might become *EMPTY*. At this stage, the capture has to be stopped for one or two cycles. Within these one or two cycles (of stopped capture), the launch clock will write into a few more locations of the FIFO – making additional data available in the FIFO – to resume the capture.

4.6.1 FIFO Basics

A *FIFO* is a 2-port memory. It has two clocks, one for *read* and one for *write*. It also has two *addresses*, one for *read* and one for *write*. A *read* happens at the location specified by *read address*, and is triggered at *read clock*. Similarly, *write* happens at the location specified by *write address*, and is triggered at *write clock*. The *read* and *write* addresses are generated by the respective counters, so that each subsequent *read* or *write* happens at the subsequent locations. The *read* and *write* addresses are typically called *read pointer* and *write pointer* respectively. Since, the memory employed is a 2-port memory, with separate address and clocks for *read* and *write*, both *read* and *write* can be independent of each other.

In the context of a *FIFO's* application in *CDC* situation, the *launch clock* is used for *write clock*. As a new data is launched, the same data is written into the *FIFO*, and, the *write pointer* is updated – for the next data to be written onto the next location. Similarly, *capture clock* is used for *read clock*. The data for capture is read from the *FIFO*, and the *read pointer* is updated – for the next data to be read from the next location. For either of the pointers, as the end of the *FIFO* is reached, it can start over from the first location of the *FIFO*.

4.6.2 Full and Empty Generation

When the *read* pointer tends to cross over the *write* pointer – it indicates that there are no valid locations to be read, which means that the FIFO is *Empty*. Similarly, when the *write* pointer tends to cross over the *read* pointer – it indicates that there are no valid locations to be written into, which means that the FIFO is *Full*. Thus, *Full* and *Empty* signals can be generated by comparing the *read* and *write* pointers of the FIFO. However, both the read and the write pointers are operating at asynchronous clocks. Hence, comparing them directly could result in failure, due to Clock Domain Crossing.

The *Empty* signal is used to control the read side of the circuitry. Thus, for generating the *Empty* signal, you should synchronize the *write pointer* to the *read clock*. Similarly, for generating the *Full* signal, you should synchronize the *read pointer* to the *write clock*. Since the *read* and *write* pointers are both address *busses*, and they both are going to be synchronized to *write* and *read* clocks respectively, so, you should use grey-counters to generate *read* and *write* pointers. Figure 4.8 shows the generation of *Full* and *Empty* signals.

It is worth noting that a FIFO would be *Full* only at the update of a *write* pointer. However, it would come out of the *Full* only at the update of a *read* pointer. The schematic shows that for the generation of *Full* signal, the *write* pointer feeds in directly to the *comparator*. That means, the generation of the *Full* signal is without

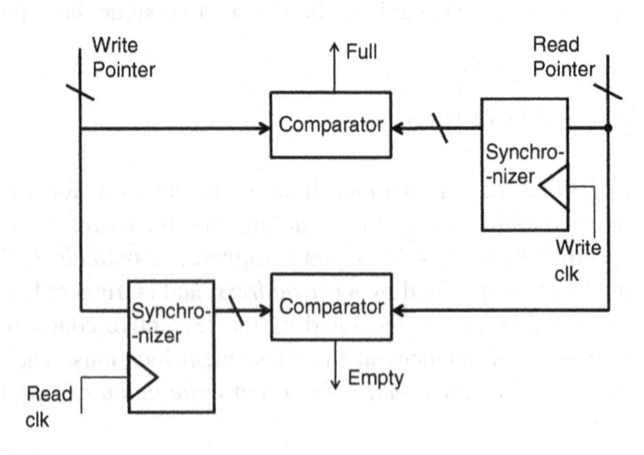

Fig. 4.8 Generation of Full and Empty

any delay. However, due to the delay (because of synchronization) on the *read* pointer side, the FIFO might take an extra cycle to deassert the *Full* signal.

Similarly, for the generation of *Empty* signal, the *read* pointer feeds in directly to the *comparator*. That means, the generation of the *Empty* signal is without any delay. However, due to the delay on the *write* pointer side, the FIFO might take an extra cycle to deassert the *Empty* signal. So, effectively, the assertion of *Full* and *Empty* happens without any delay. But, the deassertion might involve an extra cycle. This means, there might be an additional wasted cycle; however, this additional wasted cycle is worth it – to prevent any failure.

If you had done the synchronization the other way, viz: for generation of *Full*, use the *read pointer* directly, and, synchronize the *write pointer*, the assertion of *Full* would be delayed – resulting in possible loss of data, due to overwrite.

4.6.3 FIFO Limitations

FIFO might not serve any useful purpose, if the launch clock is faster (even marginally) than the capture clock, and, if the data is launched at each edge. If the data launch rate is higher than the capture rate, then, after a while, the FIFO would become *Full*. After that, the data can be written into – only after a data has been read. So, effectively, the slower clock is going to dictate the speed of data-transmission, and, the FIFO would be almost always *Full*. As soon as *Full* is deasserted, the launch clock would put another data into it – making it *Full* again. Thus, FIFO is useful only if on a long term basis, the data launch rate is not necessarily faster than the data consumption rate, even though, there might be durations where the instantaneous launch rate could exceed the capture rate.

4.7 Other Reliability Concerns

Even though you have synchronized individual bits and busses, and, have taken care of data-loss, there could still be some reliability issues, which need to be guarded against. Figure 4.9a shows a signal coming from domain *clk1* and is needed at

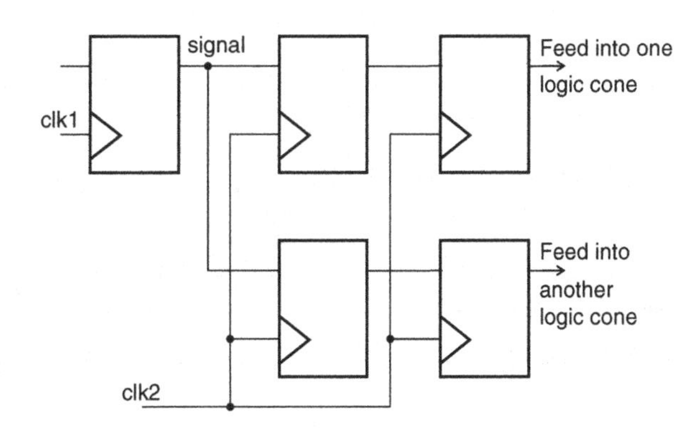

Fig. 4.9a Same signal synchronized twice – independently

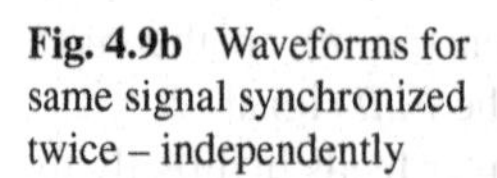

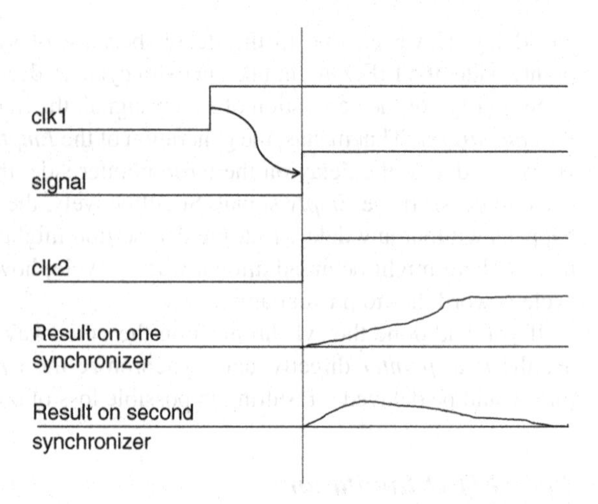

Fig. 4.9b Waveforms for same signal synchronized twice – independently

multiple places in domain *clk2*. Here, it has been synchronized multiple times – in *clk2* domain.

As can be seen in Fig 4.9b, the same signal has stabilized to two different values – in the two independent synchronizations. This might happen because once the flops have entered into a metastable state, each of them will resolve independently – which could be to two different logic levels. Thus, the same signal is being read differently – in different parts of the destination domain. So, once again, there is a loss of reliability, even though, the signal is synchronized individually – at all its points of usage.

So, it is very important that you should *never* synchronize the same signal more than once – in the *same destination domain*. If the signal is required to be used at multiple places in the destination domain, you should synchronize it only once – and then use it at as many places as required. Figure 4.10 shows the right way.

An extension of this same concept shows you should not allow to reconverge two or more signals synchronized independently. The resultant after the reconvergence could be totally different from the original combination. If a combination of multiple

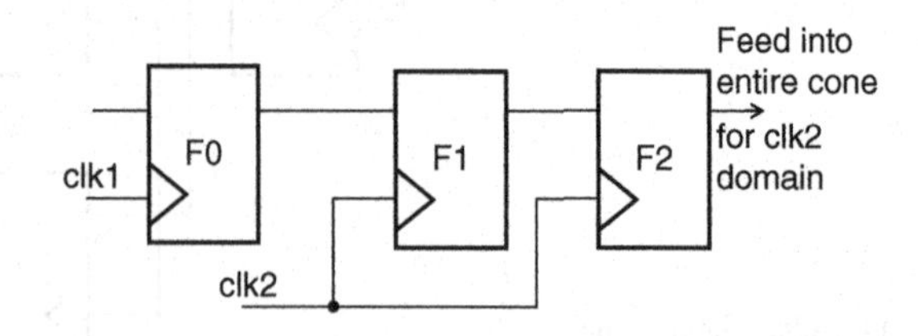

Fig. 4.10 Synchronize only once

signals is needed, they should be combined together in the source domain, and the resultant output should be synchronized to the destination domain.

4.8 Catching CDC

4.8.1 Using STA

CDC can be caught easily during *STA*. Most CDC will result in a timing-violation. But, usually you must not rely solely on STA tools for catching CDC. First of all, STA is done on a gate level netlist. That means, you can catch CDC issues only after your synthesis is done. Any problem here could mean a change in RTL – which means – redoing the functional verification effort – which is a huge amount of wasted effort. But, the most important problem is that even though an STA might report a CDC, it would still not report whether the crossing is synchronized properly. In most of the complex designs of today, there are many clocks, and, CDC is always present. What you might be interested in is not just knowing about the presence of a CDC; that is almost a given. You have to be concerned about having taken adequate safeguards to ensure a reliable operation – even when CDCs are present. And, STA analysis can only check for the presence of CDCs – and is not useful in being able to further analyze those CDCs in terms of being safe (due to adequate precautionary measures) or unsafe.

Besides, because CDCs result in timing failures, hence, the paths involved in CDC are excluded from STA (details in Chapter 7) anyways. And, so, the actual STA analysis would not report even the CDC itself – because of these exceptions being provided by the user.

4.8.2 Using Simulation

CDC might also be caught during simulation. If you are doing a *FTGS*, there might be a setup or hold violation – at the place of crossing. However, this might mean once again, waiting till gate level netlist is available. Also, the simulation only tells the occurrence of a setup or hold violation. It is up to you to correlate this timing violation with CDC being the underlying reason.

Some designers and tools are improvising the RTL simulation to catch the CDC issues at the RTL itself. In the improvisation, a random value is injected whenever the data changes at the point of CDC. This random value *mimics* the metastability at the point of CDC. Now, the simulation can be observed to see, if the circuit is able to behave properly as well as recover in the presence of a random value. However, the biggest issue with any simulation based technique is the uncertainty around ability to cover all possible situations. And, more importantly, the random value has to be injected at the points of CDC. So, this technique also depends on somebody else – to identify the point of CDC. The simulation does not help in identifying the point of CDC. The improvisation can help only in validating the ability of the design to recover from metastability.

4.8.3 Using Rule Checkers

Since the middle of the first decade of 2000, some *rule checkers* have started catching CDCs. These rule checkers can identify CDC as well as whether or not the crossings are synchronized appropriately. They are able to check whether proper precautions are taken to prevent against data-loss etc. Some of these rule checkers can work both on RTL as well as on netlist. Most rule checkers expect the users to provide certain information – in order to do a correct analysis. In the absence of correct user setup, the rule checkers might identify more issues than are actually present in the design. This is on the opposite end of the spectrum. For example, simulation based techniques might miss certain real issues – if the user input is not complete (*non-exhaustive* vector set).

The general trend is to use static rule checkers for sure (because, it does not depend on user specified vectors) – which are expected to be exhaustive checks. Several designers use a combination of both Rule Checkers as well as simulation based techniques. They use rule-checkers to identify all the crossings, which are then used for simulation based techniques – to inject random values. Also, for some complex synchronization schemes, the rule checkers might dump the corresponding assertions – which you can then feed to the simulations.

4.9 Domain Revisited

In Section 4.1, you got a brief idea of the phrase *clock domain* – in the context of this chapter. Now, look at *clock domain* a bit more closely to understand it slightly better. For any given pair of clocks, if the phase relationship between their active edges can not be determined *predictably*, the two clocks are considered to be in separate domains. And, if the phase relationship between the edges can be determined *predictably* – then the two clocks are considered to be in the same domain. Consider a few examples.

Say, there are two clocks with a period of *10* and *13*. As shown in Section 4.1, their active edges keep having different phase relationship each time. Hence, they are in two different domains. Thus, anytime there are two clocks, such that the LCM of their clock periods is a large value, or, is a product of the two periods – the two clocks should be considered to be in two different domains.

Now say, there are two clocks with a period of *2* and *4*. Assuming, edge alignment of these two clocks; for each active edge of the clock with period = *2*, the other clock (with period = *4*) has either its active edge, or, has its inactive edge. Thus, here, the phase relationship is predictable. There are two possibilities – but, the relationship is still predictable (to be within one of the two possibilities). Hence, these two clocks are in the same domain. Assuming, edges of these two clocks are not even aligned; rather they are shifted. Still, the phase relationship is determinable predictably (to be one of the two possibilities). Thus, even if the edges are shifted, they are still in the same domain.

The STA is anyways going to take the worst of the two possible situations. If the STA passes (without exceptions and with the correct clock definitions), you do not run the risk of reliability problems due to CDC. This might create an impression – if two clocks are such that the period of one clock is an integer multiple of the other (including the same frequency), they might be considered to be in the same domain, and, any exchange of data amongst these need not be considered as a CDC. A clean STA report (without exceptions) is good enough indication that the device is not going to have issues with CDC. In general, this might be true – barring just a few situations.

Now, consider two clocks which have the *same period*. It might appear that these two clocks are in the same domain – because they will have a constant phase relationship, and hence, given the edge of one of these clocks, the edge of the other two clocks would be predictably determinable. However, what if these two clocks are being generated from two totally independent sources? It is true that – once powered up, whatever phase relationship is determined between the edges of the clocks, the same phase relationship will continue to exist (leaving aside minor variations due to source jitter) across all subsequent edges also. But, when the device is powered again, this time, the two sources might exhibit a different phase relationship.

So, even though the two clocks have the same period – and are apparently exhibiting a predictable phase relationship, this relationship is predictable only during this one session of power-up. During subsequent power-ups of the device – a different phase relationship might get exhibited. Thus, the phase relationship is not predictable across power-ups. These situations should also be considered to be different domains. Thus, domain is not just about looking at period or frequency relationship. It is also about understanding the relationship between the sources of the interacting clocks. If the sources are independent of each other, the clocks are in different domains – irrespective of the clock period being same, integer multiples or totally unrelated to each other.

Actually, these situations, where, the clocks might appear to be having a predictable phase relationship (because of same or related periods), but, the sources are actually independent are very risky. STA analysis might also be done based on a constant phase relationship, which will show the STA reports to be clean, while, in reality – the device might fail. Similarly, *FTGS* will also not mimic the variation in phase relationship during different power-up sessions. This is another reason, why rule checkers based mechanisms for catching CDC related problems are more popular.

Chapter 5
Power

CMOS circuits were always considered to be consuming lower power compared to most other semiconductor technologies. However, since mid-90 s, Power became a mainstream focus area even for CMOS based VLSI circuits. Even at that time, the maximum focus was on accurate measurement and estimation at the design stage itself. It took another few years, before Power Reduction also became a main-stream activity.

5.1 Importance of Low Power

Some of the main reasons for Power Analysis, Estimation and Reduction to become important are explained in following sub-sections.

5.1.1 Increasing Device Density

Transistor geometries are shrinking and more and more functionality is being packed in a design. So, now, we have more functionality being packed into smaller areas, thereby increasing the gate-density (gates per unit area). Thus, there is more heat being generated within the same area, causing the junction temperatures to increase. We have already seen in Chapter 3 (Section 3.3.2.4), that the device delay (performance) deteriorates with an increase in junction temperature. Besides, higher junction temperature reduces the reliability of the design. The need to maintain the junction temperature puts a requirement to put higher effort in dissipation, either through heat-sinks or costlier packaging. Hence, there is an interest in reducing power dissipation.

5.1.2 Increasing Speed

As transistors are becoming smaller, they are also becoming faster. We will see in Section 5.3.1 that switching of signals means energy consumption. So, increasing

S. Churiwala, S. Garg, *Principles of VLSI RTL Design*,
DOI 10.1007/978-1-4419-9296-3_5, © Springer Science+Business Media, LLC 2011

frequency (faster devices) means energy consumption in a lesser duration, thereby increasing power. This also raises the junction temperature, thereby causing issues of reliability and performance.

5.1.3 Battery Life

Starting mid-90 s there has been suddenly an advent of portable products. These portable products are really hand-held, battery operated devices. These are different from earlier generation of portable products, which were truck-mounted and powered by generator sets. Most of these current generation portable products are in the consumer or communication segments. For these hand-held battery operated devices, the battery-life is very important feature. Hence, in order to provide a longer battery life, there is an interest in reducing power consumption.

5.1.4 Green Concerns

Earlier, concerns around environment and its sustainability was mostly a matter of concern for environmentalists only. But again starting mid-90 s, the environment-conscious movement took on a stronger and worldwide focus. Products are being characterized against their carbon foot-print, energy efficiencies etc. Hence, designers are putting in more effort to reduce power, in order to conform to these movements and the legislation that has resulted.

5.1.5 User Experience

There has been an advent of applications such as gaming, video-streaming etc. which are all about enhanced user-experience. These user-experiences need very high performance. High performance means higher power consumption. However, when the user is using the same device for other applications which do not need this high performance, it should be able to conserve battery for longer durations. So, there are multi-application devices, where it is OK to consume high power for applications that need high performance (e.g. games), but on the other hand, power should be saved on applications where performance is not that critical (e.g., messaging).

5.2 Causes of Power Dissipation

In this chapter, we have used Power and Energy interchangeably, specially, where we talk about an instantaneous consumption of energy. Instantaneous energy consumption – when aggregated over the duration means power consumption. For example, consider average power to be computed over a period of 5 min. In one situation, there is a dissipation of 1 mW for 5 s. For the remaining duration, there is no power consumption. In another situation, there is a dissipation of 1 mW for 30 s. For the remaining duration, there is no power consumption. So, while, the instantaneous energy used for both these circuits is the same (when they are consuming power), the second circuit has consumed 6 times more total energy, over the entire duration. And, when aggregated over the full 5 min duration, the second circuit has consumed more power.

Also, for the concerns which mainly drive power analysis and reduction – the main interest has been total energy, rather than power. Hence, there are portions in the chapter, where, we refer to charge (rather than current) or energy (rather than power). Once you have a good understanding of the concepts, you can easily convert one to another. We have used energy, power, charge, current whichever makes most natural and intuitive sense – in the context of that specific discussion.

As a first step, look at the causes and components of Power Dissipation in a CMOS VLSI circuit. Consider the circuit for CMOS inverter shown in Fig. 5.1.

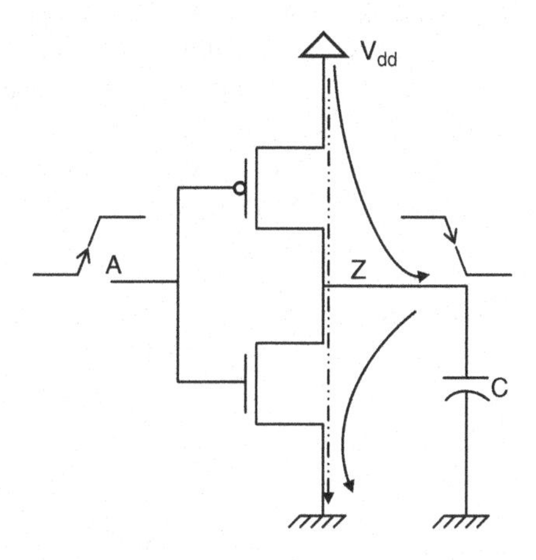

Fig. 5.1 Dissipation of power in CMOS circuits

Consider a transition from Low to High on A. Initially, A is Low. The P-transistor is ON. Current flows from V_{dd} through the P-transistor to the output – Z and is used to charge the capacitor C. By the time the capacitor is fully charged it has stored energy equivalent to $\frac{1}{2}CV^2$ on it.

As A starts rising, the P-transistor starts turning OFF, and, the N-transistor starts turning ON. There is a short duration (typically, when A is around mid-way of V_{dd}), when both P and N transistors are partially ON. During this duration, current flows from V_{DD} through the P-transistor through the N-transistor and to the *Ground*. This flow of current is shown as a broken line in Fig. 5.1. Once A has reached High, the P-transistor is completely OFF, and, the N-transistor is completely ON. The capacitor C which was charged to V_{dd} now starts getting discharged through the N-transistor. The charge stored in the capacitor is now discharged.

Even when a transistor is OFF, there is always a very small amount of leakage current flowing through its reverse biased p-n junction. So, consider when A is at High, and, C has been totally discharged, still there will be a leakage current flowing through the P-transistor. The above sets of activities pretty much resemble the entire Power dissipation cycle in any CMOS VLSI circuit.

The consumption of power due to repeated charging and discharging of the Capacitor is called Switching Power. The consumption of power due to instantaneous path established between V_{dd} and *Ground* when A was transitioning is called Short Circuit Power or Internal Power. The consumption of power due to leakage currents flowing through OFF transistors (i.e. through reverse biased junctions) is called Leakage Power.

The Capacitor mentioned in the context of Switching Power is not an explicit capacitor inserted in a design. Rather, it refers to the effective capacitive load – exhibited by the input of the next stage that is being driven by this circuit plus capacitive component of wire-load.

While the schematic shown is for a CMOS inverter, you can consider this to be a representation of any generic CMOS ASIC gate. For any gate, at its output stage, there will be one or more P-transistors connected in series/parallel, which will establish a path to charge the capacitor – in order to take the gate output to High. Similarly, there will be one or more N-transistors which will establish a path to discharge the capacitor – in order to take the gate output to Low. Consider an AND CMOS cell as shown in Fig. 5.2 as an example of a generic ASIC cell. Though, we have chosen to show an AND cell, the dotted line box represents the boundary of any generic CMOS ASIC cell.

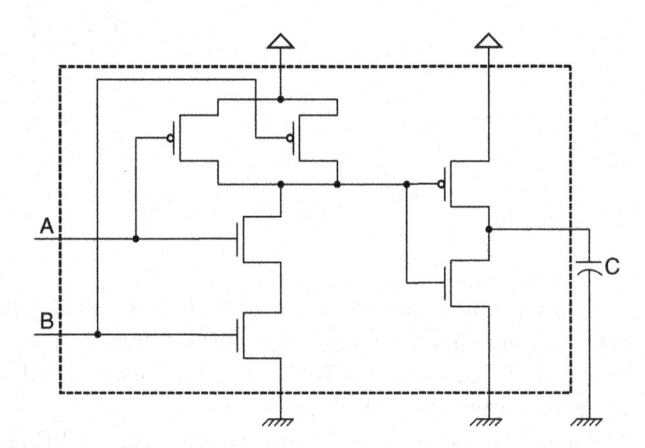

Fig. 5.2 AND cell as an example of generic CMOS cell

If we look specifically in the context of this cell, we can see power consumption as:

Charging and discharging of capacitor *C* as the gate output goes Low-to-High and High-to-Low. This might be called as Load Power or even Switching Power.

As inputs A and B toggle, there will be some power consumption inside the cell. This power consumption will take place mainly due to:

- Flow of current through instantaneous short-circuit paths (referred earlier as Short Circuit Power)
- Repeated charging and discharging of nodes INSIDE the cell. These nodes are all exhibiting capacitive load themselves.

These two together maybe called *Internal Power* – as this refers to the component of power that is dissipated internal to the cell. Internal Power and Switching Power are together also called *active power*, because, they are caused due to activity in the circuit.

Leakage Power refers to the current flow from V_{DD} to *Ground* inside the cell, through reverse biased junctions. In the context of ASIC cell based design, the entire Leakage Power is within the cells.

Depending upon the application, different components of power could be of more importance. Typically, for devices with high amount of activity, the leakage power could be very less – compared to active power. However, for devices, which are usually ON but inactive (e.g. Cell Phones, Alarm Circuits etc.), leakage power becomes significant. In such applications, leakage power considerations are also important – for the sake of longer battery-life.

5.3 Factors Affecting Power

Now, that you have a better understanding of exactly how and where does the power get consumed, consider the factors which affect Power.

5.3.1 Switching Activity

Higher is the times there is a transition at an input, the greater will be the times when the instantaneous short circuit is established. And, during each such instantaneous short-circuit, there is a short-circuit power. Similarly, higher are the times, that the output switches, the higher is the number of charging and discharging of the load capacitance. So, the number of transitions or switching will impact both Switching Power and Short Circuit Power (or, Load Power and Internal Power – in the context of an ASIC cell).

5.3.2 Capacitive Load

During each charging, the charge stored is directly proportional to the capacitive load. This same amount is going to be discharged. So, during each cycle of the load switching, the amount of charge transferred from V_{dd} to *Ground* (via two transitions of the Load capacitor) is dependent on the load value. Higher the load value, higher is the Switching Power.

5.3.3 Supply Voltage

Supply voltage impacts the power consumption in several ways. More importantly, it impacts all components of Power Consumption. The charge stored on the capacitive load is proportional to the square of supply voltage. So, higher supply voltage means greater charge stored on the capacitor during charging. This in turn means, flow of higher amount of charge from V_{dd} to *Ground* (via, load capacitor). So, switching power is proportional to the square of supply voltage.

When an instantaneous short circuit is established due to inputs transitioning, the transistors taking part in this short-circuit path behave like resistive components. Hence, the short circuit current flowing through these transistors would depend on the supply voltage. Higher value of V_{dd} would mean a higher current flow. So, Short Circuit Power is proportional to the square of the Supply Voltage. The leakage current flowing through the reverse biased junctions (i.e. OFF transistors) is dependent on the Supply Voltage. Higher supply voltage would mean higher leakage currents.

5.3.4 Transition Rate

If a signal is changing slowly, it spends longer time around $\frac{V_{dd}}{2}$ area. That means, the instantaneous short circuit path remains established for longer durations. This in turn means, the current flow from V_{DD} to *Ground* occurs for a longer duration.

5.3.5 Device Characteristics

Some people refer this as Transistor Geometries. We have used the Device Characteristics to also include several other aspects of fabrication process, which finally impact the I-R characteristics of a device. Devices having higher resistance would cause lower instantaneous short-circuit current. Similarly, device characteristics will also impact the leakage current flow.

5.3.6 Device State

Referring to Fig. 5.1, when *A* is at Low, the P-transistor is ON, and the leakage power is dictated by the leakage current through the off N-transistor. Similarly, when *A* is High, the N-transistor is ON, and the leakage power is dictated by the leakage current through the OFF P-transistor. The amount of leakage current through both these transistors would be different. So, in steady-stage, the amount of leakage current would depend on whether *A* is at High or at Low.

Now that you know all the factors that impact power, the issue of Power reduction is mostly limited to controlling one or more of these factors. In the next few sections, you will see how to control some of these factors in order to reduce power consumption.

5.4 Switching Activity

Section 5.3.1 showed that by reducing Switching Activity, you can reduce Short Circuit Power as well as Switching Power. The sections below explain some of the common techniques used for reducing Switching Activity.

5.4.1 Shifter Instead of Multipliers

In an *XOR* gate, any transition at any of the inputs will always cause a transition at the output. Compare that with gates like *AND, NAND, OR, NOR* etc. Not all the transitions at an input will result in a transition at the output. At least, some of the transitions will not reach the output, because the values get blocked by the other input (*0* for *AND, NAND* and *1* for *OR, NOR*). So, in general, an *XOR* gate is more power hungry, because it transmits all transitions at inputs into a transition at output. Multipliers (and adders) are made of a large number of *XOR*s. So, even a single transition at any input will result in a lot of transitions internal to the multipliers. There are certain multiplication values, where, the same result can be achieved using another operation, e.g. shifter instead of multiply by 2, 4, 8 etc. Where possible, the multiplier should be replaced by a Left Shifter. Similarly, some divisions (by 2, 4, 8 etc.) can be replaced by a Right Shifter.

5.4.2 Operand Isolation

Figure 5.3a shows a large adder, which adds 2 signals – each of 32 bits.

Fig. 5.3a Addition of 2 data buses

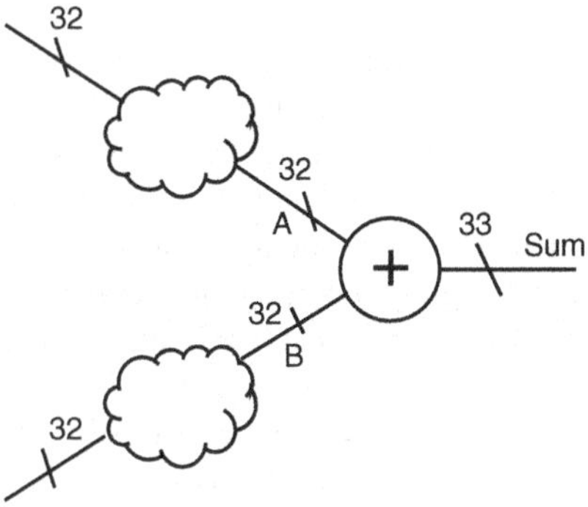

All the 32 bits of the busses *A* or *B* will not arrive at the adder at the same instant. These different bits could be arriving through different combinational gates, or, through different routing tracks. As each signal at the input of the adder changes, the adder reevaluates and the output switches. This activity starts with the arrival of the earliest of the 64 bits (32 bits each of each *A* and *B*), and, continues till all the bits have arrived. However, only the final addition value is of interest. All the intermediate results were of no use, but, they consumed power. Arithmetic operators such as Adders and Multipliers have lots of *XOR*s. So, any transition at the inputs will reach the output – thus causing more switching activity down the circuit. So, the aim is to reduce the number of activities at the inputs of these operators. This can be done by latching all the inputs, and, then, feeding the latch output directly to the adder. This ensures that all the bits are presented to the operator at almost the same time. This prevents the operator to evaluate multiple times. Hence, a lot of spurious switching can be avoided. Figure 5.3b shows the modified circuit.

Fig. 5.3b Latching operands

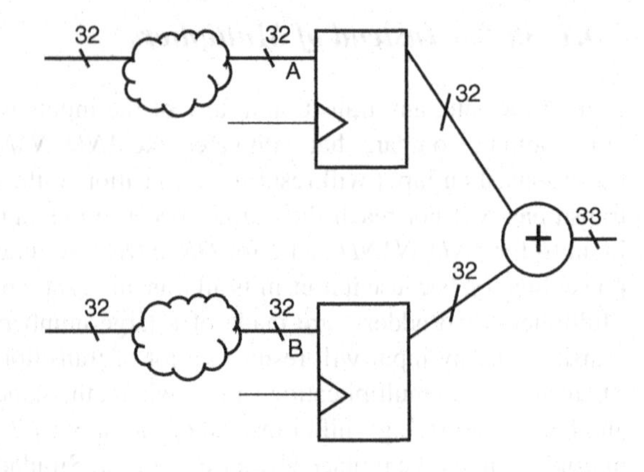

Since all the operands are now coming directly from the registers, they will reach the operator at almost the same time. The cost of using this technique is some area-penalty, as you have to add these additional registers. Also, there is an increase (by one clock cycle) in latency.

5.4.3 Avoid Comparison

Comparator operations also involve XORs. It you have to compare large data-busses, avoid comparing all the bits in one go. If there is a change in any of the inputs to the comparator, it is possible that the final output remains unchanged. However, there might still be lot of intermediate transitions – which are effectively glitch. These intermediate transitions also consume power. This power consumption is called glitch power. Glitch power is totally wasteful, as these transitions do not contribute to the final functionality. Hence, wherever possible, you should try to prevent glitches. In the context of a comparator, statistically, comparing just the MSB would be able to give the results in 50% of the cases. Hence, the lower bits should be involved in the comparison, only if the MSB comparison is non-diagnostic. Figure 5.4a shows a typical comparison of two large busses (32 bits each) – which involves 32 *XOR*s.

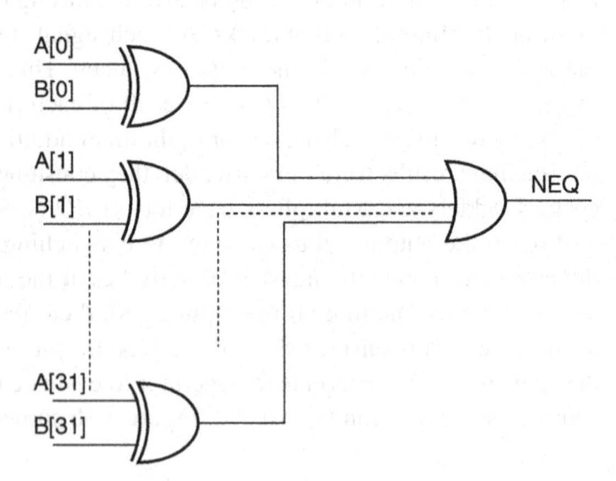

Fig. 5.4a Typical comparison of data buses

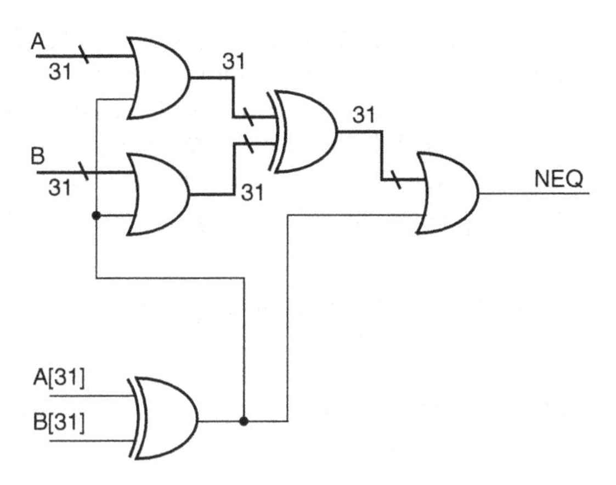

Fig. 5.4b Compare MSB; then lower bits – if needed

Figure 5.4b shows a modified implementation, where, the MSB comparison is done first. If the MSB alone is sufficient, the other bits are prevented from taking part in the comparison. Only if the MSB is not able to take a decision, the lower bits are allowed to go to the *XOR* gates – for their comparison.

Care should be taken that the lower level bits of the busses should reach the XOR, only after the MSBs have been evaluated. Else, if the lower level bits reach the respective XORs, they would anyways create the glitch, and there will be no advantage of pre-computing the MSB. The same concept can be further applied recursively – for the comparison of the lower 31 bits also. That is, do the comparison for the MSB among these lower 31 bits, and, only if this is non-diagnostic, allow the comparison of the still lower 30 bits. And, so on.

The cost of doing this modification is:

- Additional area required to implement additional circuitry.
- The timing characteristics of the paths get modified.
- For the lower bits, an additional gate is now introduced. If these were already in the critical path, they will get further delayed.
- For the MSB, there is an additional load on the XOR gate, which will deteriorate the timing on this path. If this bit was on the critical path, it will further deteriorate the timing.
- Every time the MSB comparison switches, it will cause additional switching power – due to additional load being seen by its output.

These above cost are to be kept in mind, while determining, the extent up to which this pre-computation (of MSB) should be done – before letting the LSBs take part in the actual comparison operation.

5.4.4 Clock Gating

Clock gating is the most popular and most effective method to reduce switching rate. Consider a positive edge triggered flop. For every negative edge on a flop's clock terminal, there is no change in the output of the flop. So, this clock edge is only consuming power (within the flop), without doing anything substantial. However, this negative edge can not be avoided. The clock signal which has gone High, has to be brought back down to Low – so that it can have the next positive edge!!!

There might be situations where the data need not be captured on the flop – even at the next positive edge of the clock. These situations include:

- When output is not being sampled. Since the output is not being used, there is no point updating its value.
- When input is not changing. Since the input has not changed, no point sampling the same input – again.

So, when there is a situation that a flop does not need to capture the new data, the clock itself can be stopped. This will prevent both negative as well as the positive edge of the clock on the flop terminal. Figures 5.5a and 5.5b show two such situation, when the clock can be gated.

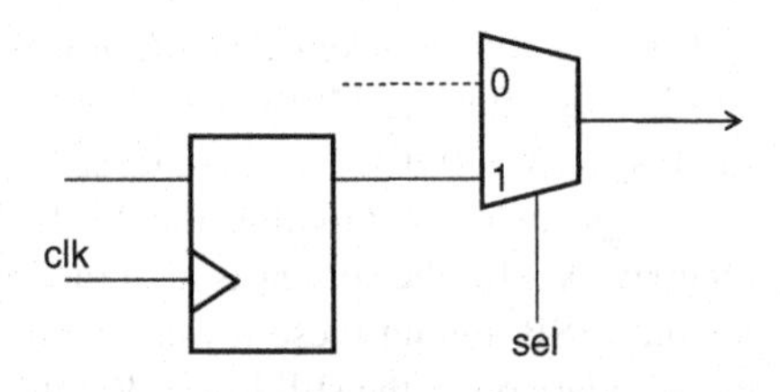

Fig. 5.5a Flop output of no use, if *sel* = *0*

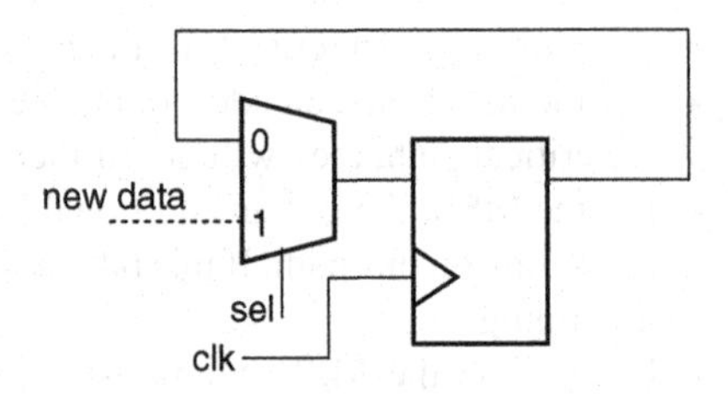

Fig. 5.5b Flop output will not change, if *sel* = *0*

Figures 5.5c and 5.5d show the corresponding clock-gated versions of the circuit. It should be easy to see, that when the flop's output is not to be used/updated, the clock will be blocked at the *AND* gate, and, not reach the flop's clock terminal. The signal which is used as the other input of the *AND* gate is usually called *enable* (or, *gate-enable*), because, it enables or disables the clock to cross the *gate* and reach the flop's clock terminal.

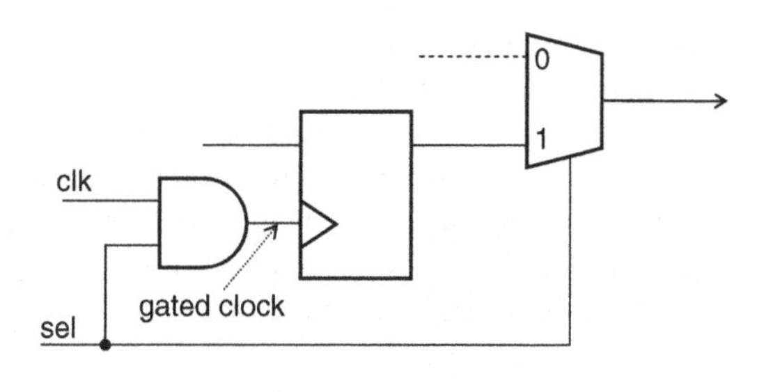

Fig. 5.5c Gated clock implementation for Fig. 5.5a

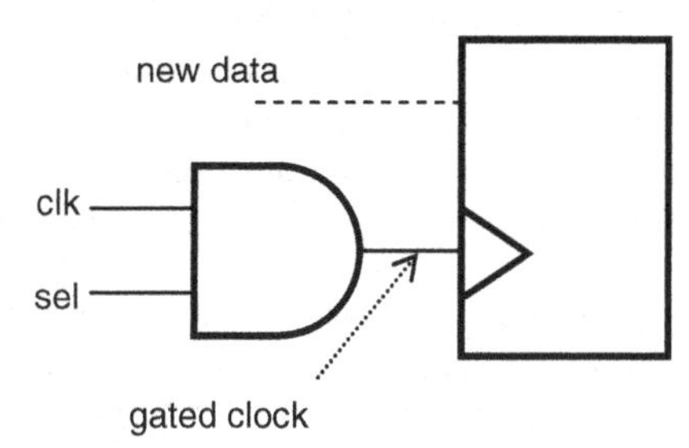

Fig. 5.5d Gated clock implementation for Fig. 5.5b

5.4.4.1 Pulse Clipping, Spurious Clocking

Figure 5.6 shows a possible waveform at the *gating* circuit.

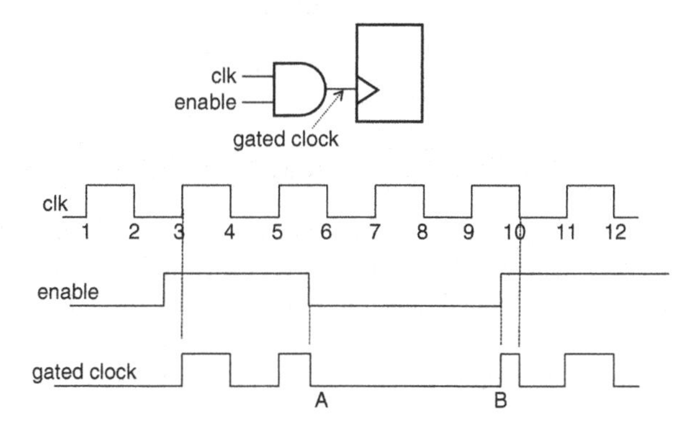

Fig. 5.6 Operation of gating circuit

It should be easily observable that because of the gating, edges 1, 2, 7, 8, 9 of the original clock have been prevented from reaching the flop's clock terminal, thus saving power within the flop. Look more closely at the point *A* on the waveform for the gated circuit. The falling edge which should have come at *6* has come earlier. Effectively, the *High* pulse of the clock has been shortened. This is called *pulse-clipping* and can result in Pulse width Violation for the flop being clocked by this gated-clock.

Look also at the point *B*. There is a positive edge at point *B* on the gated clock. This is not where a clock should have come. This is called *spurious clocking*. This can result in two possible problems:

- By the time that this flop has received the spurious clock, its D could also be getting updated with the new values. This new value (which might not yet be stable, but, just a glitch) could get captured in the flop. In terms of STA, this is a situation of a Hold-Time problem on this flop.
- Since this flop updates its new value at B, the next stage flop does not get one complete cycle for the value to reach there. That is, there could be a setup-time problem on the flop at the next stage.

To prevent these problems of Pulse-Clipping or Spurious Clocking, you need to ensure that the *enable* signal is received on the *gate* only when the clock is Low. That way, the *enable* itself will not cause any transition at the output of the gate, but, will be able to control the transitions of the clock. Figure 5.7 shows an implementation to achieve the same, and, the corresponding waveforms.

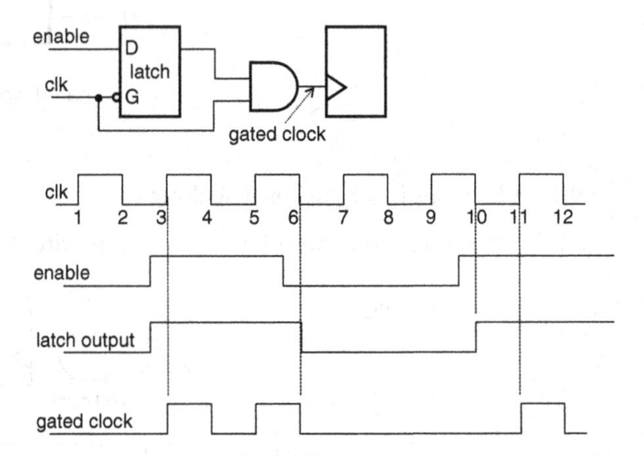

Fig. 5.7 Preventing pulse clipping and spurious clocking

The transparent-when-Low latch will block enable's transitions from reaching the *AND* gate, till the clock has become Low. Thus, irrespective of when the *enable* is actually generated, this transparent-when-low latch will ensure that the *AND* gate will see a transition on *enable*, only when the clock is low. This will prevent any spurious clocking or pulse-clipping on the *gated_clock*.

5.4.4.2 Integrated Clock Gating Cell

Assume, due to differential routing delays, it so happens that the clock to the *AND* gate path is faster than the clock-to-latch path. Figure 5.8a shows the waveforms for this situation. The dashed wire for clk connected to G terminal of the latch denotes that the actual routing is different (longer) than shown in the figure. Longer the wire more is the delay.

This shows, how a spurious clocking can still be caused – due to this differential delay in clock paths. Figure 5.8b shows the waveforms for the reverse situation, when due to differential delays, the clock to *AND* gate path is slower than the clock-to-latch path.

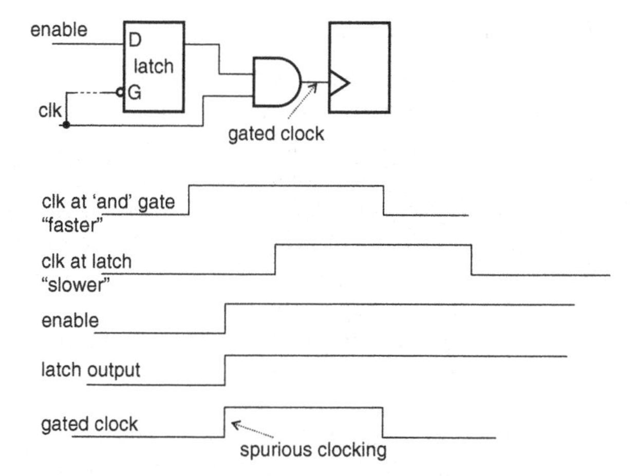

Fig. 5.8a Spurious clocking due to differential delay

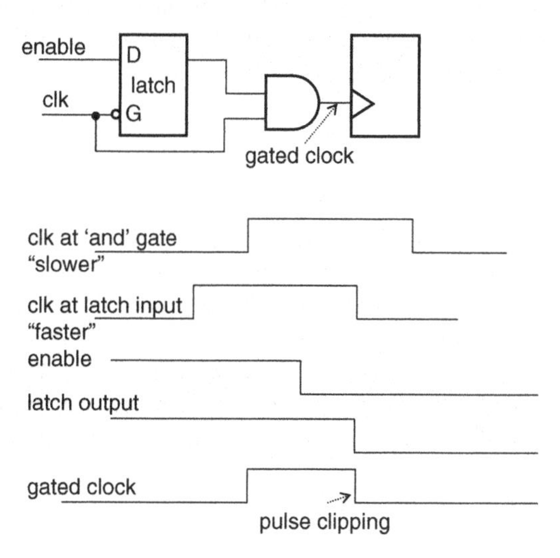

Fig. 5.8b Pulse clipping due to differential delay

In these diagrams, the differential delay has been magnified dramatically to bring out the implication clearly. However, for these issues to manifest, the delay differential need not necessarily be this dramatic. Cell Placement and routings can easily create sufficient differential for these problems to be exhibited.

In order to prevent these delay differentials, the latch and the AND gate are put together in a cell, and, their path-delays are well balanced. This entire circuitry is together treated as a single cell, called *Integrated Clock Gating cell* (or, *ICG cell*). So, a placement tool can never change the differential placement of the *Latch* with respect to the *AND* gate. And, since the wire-routing within this ICG cell is already done, a router cannot insert any additional differential.

Now, you have a clock-gating mechanism that does not create any functional risks – as long as you keep the functionality to be logically correct.

5.4.4.3 Cost of Clock Gating

When you do clock-gating, you take up some silicon area – in order to put an additional latch and the *AND* gate. While, you are saving the clock transitions from reaching the flop's clock terminals, these transitions are now reaching the *AND* gate and the latch, which were not there earlier!! Also, there are some transitions which do reach the flop's clock terminal. These transitions also are reaching these additional gates inserted. So, there will be some additional power consumption, due to this extra circuit.

Usually, a clock-gating is not done on an individual flop-by-flop basis. Rather, if one gating signal can impact many flops, then, it might be worthwhile to do the clock gating. That ensures that the additional power consumed by the gating logic is much lower, compared to the power saved on so many flops. Also, if the gate is expected to be ON for most of the duration, then, anyways, the flops will see almost entire activity on their clock, and, there will be unnecessarily additional activity on the gating circuit. So, clock gating makes sense only if the *enable* signal is expected to be OFF for a significant duration, so that a significant number of clock pulses can be eaten up – at the *ICG cell.*

5.4.4.4 Gating Location

Wherever possible, gating should be done as close to the clock source as possible. Clock lines have very high drive buffers, which consume a lot of power. So, the closer to the source you do the gating, the more power you can save. Figure 5.9 shows a typical clock distribution system. If you can gate the clock at *A* itself, it will save transitions through the entire network. If you can do the gating at *B*, it will save transitions on *I3* buffer, but, the transitions on *I1* and other branches would still be happening. And, if we do the gating at *C*, then, even *I3* will see the transitions.

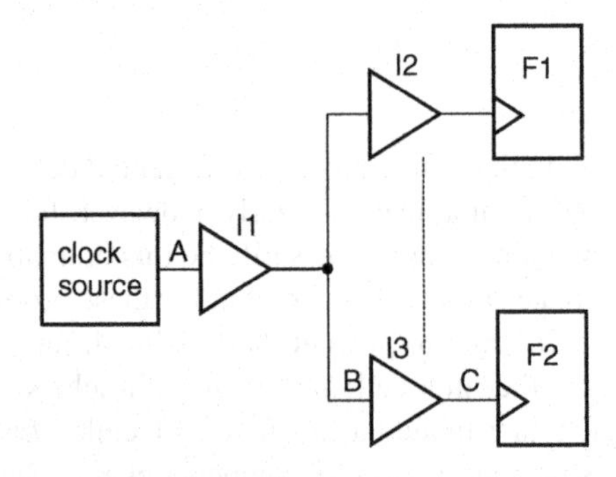

Fig. 5.9 Preferred location of clock gating

Though, the aim might be to gate as close to the source as possible, sometimes, it might not be possible, because, the closer you are to the source, the higher is the cone of impacted flops. For example, if you do the gating at *A*, then, both flops *F1* and *F2* get impacted. But, if we do the gating at *B* (or *C*), *F1* is unaffected, and, only *F2* is affected. So, by staying closer to the flop (i.e. away from the clock source), you can do a much finer tuning on the choice of the gating signal, so that maximum possible clock pulses can be eaten.

Sometimes, a mix can also be used. Figure 5.10a–c shows three ways of doing such gating. Say, flop *F1* needs to receive new data only when $A = 1$ and $B = 1$; and flop *F2* needs to receive new data only when $A = 1$ and $C = 1$;

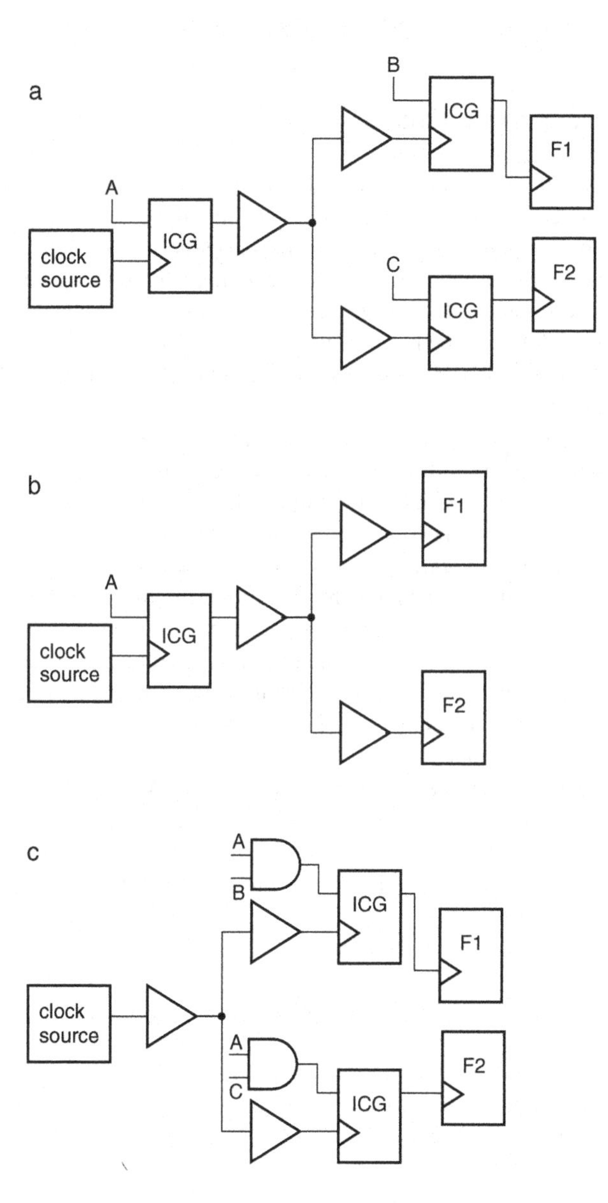

Fig. 5.10 Some alternatives for location of clock gating

The actual choice will have to depend upon:

- The number of flops impacted
- The overall duration – for which the gating circuits block the clocks.

If *A* is expected to be at *0* for most of the duration, the scheme shown in Fig. 5.10a or b could be of use. If *B* and *C* are expected to be at *0* for most of the duration, the scheme shown in Fig. 5.10a or c could be of use. These 3 alternative locations are presented as representative only. There are several more alternatives possible.

5.5 Supply Voltage

Section 5.3.3 showed that all components of power are related to supply voltage. Hence, a simple technique could be to reduce the Supply Voltage. This technique is able to make a huge impact on power reduction and hence is very common. The only major requirement is that the technology should support the devices operating at a lower voltage. In fact, the progress on this front has been phenomenal. During early 90 s, 5 V was the norm in VLSI designs. But, today, ~1 V is the norm. Though, the method is as simple as this, this has a huge direct impact of deteriorating the device performance.

So, you might have to use multiple voltages. Components on the critical path need to work at higher voltage. While, lower voltage may be used for components lying on path with enough timing slack. *Voltage domain* refers to the portion of the circuit which operates at a specific voltage level. In the world of physical design, the term used is *voltage island* – because, circuitry operating at the same voltage levels are placed in immediate physical proximity to each other – so that they all can share the same power grid. So, it appears to be an island of cells which operate at a specific voltage level, and, surrounded by cells operating at other voltage levels!!!

So, for a design having multiple voltage domains, you would have signals going from a lower voltage to a higher voltage level, and, vice-versa. This can have an impact on noise-margin, thus, reducing the reliability. Figure 5.11a shows a signal generated with 0.8 V supply going to a 1.1 V supply.

In the 0.8 V domain, the switching threshold (*at* $\frac{1}{2}V_{dd}$) is 0.4 V, which means a noise-margin of 0.4 V. In the 1.1 V domain, the switching threshold is at 0.65 V. So, when a High Signal at 0.8 V signal reaches this 1.1 V domain, the noise-margin is

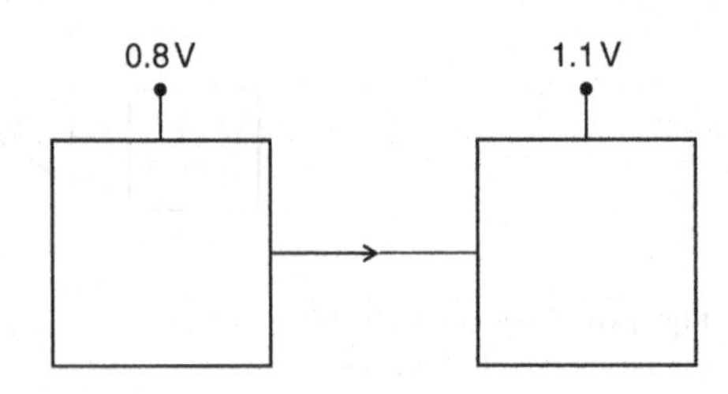

Fig. 5.11a Signal going from a lower voltage to a higher voltage

reduced to just 0.15 V (0.8 – 0.65) Hence, there is a need to put in a level shifter when a signal moves from one voltage to another voltage. Figure 5.11b shows the signal moving from one voltage to another – through a level shifter.

Fig. 5.11b Signal crossing voltage boundary through level shifter

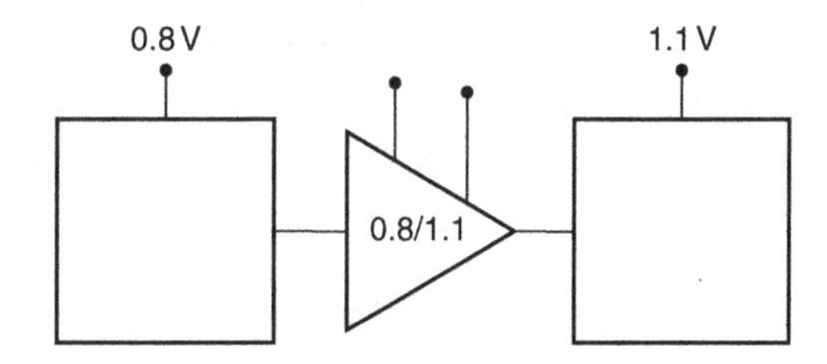

As should be visible, for each combination of source and destination voltage levels a different kind of shifter is used. For example, the shifter used for 0.8 V-to-1.1 V is different from the one used for 1.1 V-to-0.8 V. The difference is not just for low to high voltage levels -vs- high to low voltage levels; rather, each unique pair needs to have its own shifter. Sometimes there might be situations of a lower voltage signal entering a module of higher voltage, but, there is no need for a level-shifter. For example module just acts as a feedthrough for the signal. Figure 5.12 shows some situations where level-shifters might be needed or not needed.

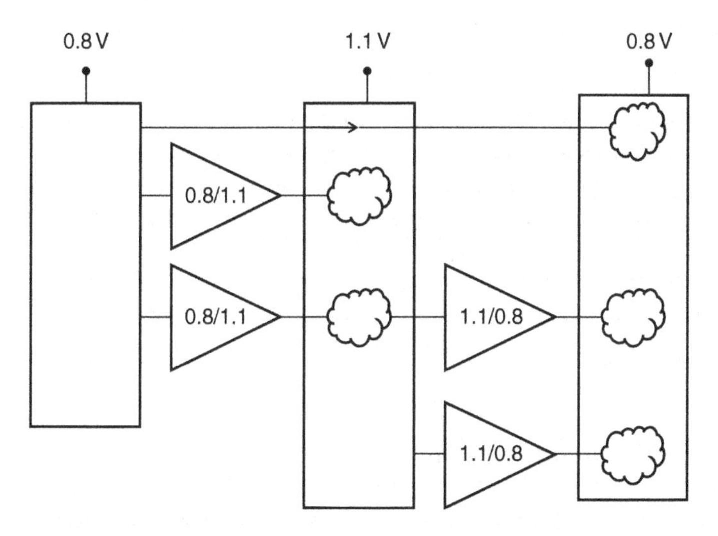

Fig. 5.12 Level shifter requirements

5.5.1 Simulation Limitation

During HDL simulation, there is no concept of distinguishing between supply voltages. Usually, designs for simulation do not even have supply and ground connectivity, either at the gate-level or at the RTL level. So, if a signal is going across different voltage levels, a simulator is not going to be impacted, because, all it sees is logic levels, and, not the absolute voltage values. Similarly, even if a level shifter is present, but, of the wrong type (i.e. one which was not supposed to be for this voltage combination), the simulator would have no way of catching and reporting this. One solution could be to revert to analog simulations like SPICE – but, that would mean too much time for simulation. A popular solution today to catch missing or incorrect level shifters is to use rule based checkers.

5.5.2 Implication on Synthesis

A level shifter inserted in the RTL (or, even netlist) might appear simply as a buffer to a synthesis tool. As it tries to do various kinds of optimizations, a synthesis tool might remove this buffer, or, replace this buffer with another buffer. In order to prevent this, there has to be a directive to the synthesis tool, not to replace/remove this buffer. In Synopsys family of synthesis tools, this might be done through *set_dont_touch* on the shifters. For other tools, the corresponding command or attribute could be different. Similarly, a level-shifter cell in the library might appear as a buffer. Hence, a synthesis tool might make use of such level-shifters, when it intends to use a buffer. Usually, the library description of these cells is expected to have a directive on these cells – so that they don't get picked up – in place of a buffer. For Synopsys .lib, this is achieved through *set_dont_use* attribute in the .lib description of the cell.

Still, without depending on the presence or absence of the attribute in the synthesis library, you should give a directive to the synthesis tool – to not pick up level shifters on its own. In Synopsys family of synthesis tools, this might be done through *set_dont_use* command in the synthesis scripts. Besides, the synthesis tool and methodology should have the support for multiple voltage libraries. When power was relatively new concerns around mid-90 s, this was a major bottleneck in the use of multiple voltage techniques. However, currently, this is not a concern, as synthesis tools and methodologies have now added support for multiple voltages.

5.5.3 Implication on Backend

Though, backend is beyond the scope of this book, however, we will briefly touch the backend implications. Finally, the cells need to be connected to their power and ground lines during *global routing* stage. Now, these tools will have to route not just one kind of Supply voltage line, but, of several kinds. And, while connecting the cells to the different power lines, it will need to ensure that each cell is connected to the right power supply. Level shifters have to be connected to two power supplies. These tools again need to ensure that the two lines connected to the level shifters are correct and are also in the right order. Routers have the ability to do some in-place optimizations through upsizing and downsizing. Upsizing and downsizing refer to making very local optimizations, wherein, a specific cell is replaced by another cell having exactly same functionality, but higher or lower drive strength. With respect to this functionality, they face the same issues as synthesis tools (mentioned in Section 5.5.2) – in terms of distinguishing between buffers and level-shifters.

So, after Place and Route, there would be a PG (Power-Ground) view of the netlist. In this view, the cells are modeled with their Power and Ground pins and those pins are also connected. For Example, a simple 2 input *AND* gate is supposed to have 3 pins – 2 inputs and one output. However, in the PG view netlist (and, the corresponding .pglib library), the same cell will have 5 pins – 2 inputs, one output, a power supply and a ground pin. This PG view of the netlist can be subjected to rule based checkers to make sure that the cells have been connected to the appropriate power lines.

5.6 Selective Shut Down

Portions of the circuit which are not in use can be selectively shut down. This approach has been in use for long – at a macro scale. The computer screen going blank when not in use for long, calculators turning OFF after defined period of inactivity are all examples of this concept. However, with Power being a major focus, the same concept is now applied at a much granular level i.e. within the VLSI circuits. Since there is no activity anyways, so, the only savings through selective shut down is for leakage power. As we know in this age of sub-micron VLSI circuits even the leakage power is a substantial portion of the total power consumption. Hence, it is useful to save it. So, this concept of selective shutdown is used in most of the VLSI ICs these days especially when the IC is for battery operated devices.

Suppose there is a portion of the circuit – which is expected to be non-active for long durations. The power supply to this portion of the circuit might be switched off, during such durations of non-activity. Figure 5.13 shows the schematic for selective shut down.

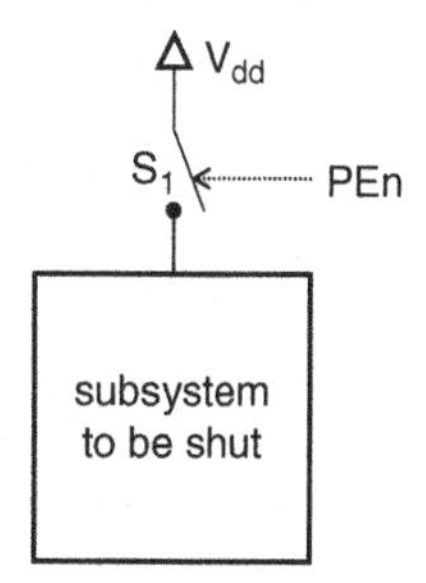

Fig. 5.13 Switch for selective shut down

The switch S_1 is usually an active device (say: low-leakage P transistor), controlled through *PEn* (Power Enable) signal. When the sub-system is supposed to remain idle, the switch would be opened, thus, breaking the power supply. Some people refer to *PowerDn* signal (which is inverse of PowerEnable).

Power Domain refers to a portion of the circuit, which operates at the same voltage level and turns ON or OFF together – by the same Power Enable signal. In the physical design world, this is called Power island. The design of such selective shut-downs requires a lot of care. The most important aspect is that when the system receives back power (to start an activity), it should be able to start from where it was when the power was shut off. Hence, all the sequential devices should be able to retain their stored values, even when they were powered down. Such flops are called *Retention Flops.*

5.6.1 Need for Isolation

Figure 5.14 shows a situation where a sub-system which is OFF is feeding into another sub-system, which is ON.

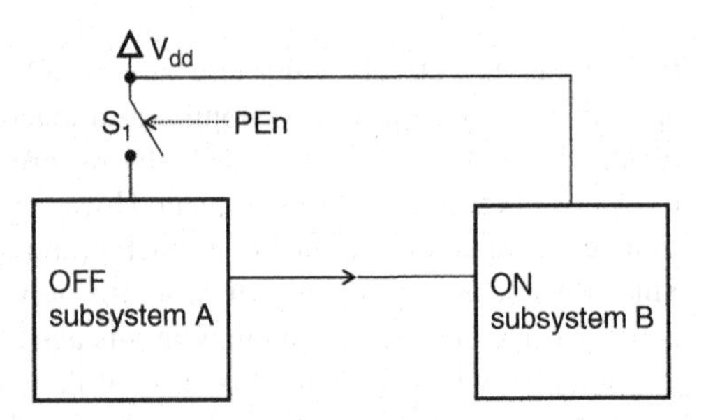

Fig. 5.14 OFF subsystem feeding into ON subsystem

This second sub-system could be ON, because it does not have a Power Switch (i.e., its Always ON), or, it has a different Power Enable, which has not turned it OFF. In either case, the subsystem *B* is expecting a signal from a sub-system that has been turned OFF. This could cause this specific input of *B* to be *floating*. A *floating* signal might neither be at 0 or at 1; rather it might lie somewhere in-between the two levels. Thus, inside subsystem *B*, both the *P* and the *N* transistors might get turned partially ON – thus, causing a direct current path from supply to the ground. This would cause the device to be burned. Thus, there is a need for an *Isolation Cell*, which isolates the OFF subsystem from the ON subsystem, and ensures that the ON subsystem does not see a *floating* input.

There are 3 kinds of Isolation Cells. Figure 5.15a shows an Isolation Cell, which sends the normal signal, when Powered, and, a *0* during Power Down. A high on *PEn* denotes normal mode of operation, and, a "0" indicates Power Down.

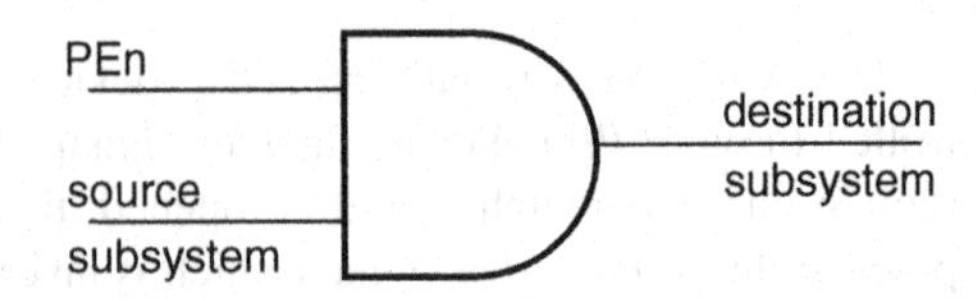

Fig. 5.15a Isolation cell to send a "0" during power-off

Figure 5.15b shows an Isolation Cell, which sends the normal signal when Powered, and a *1* during Power Down.

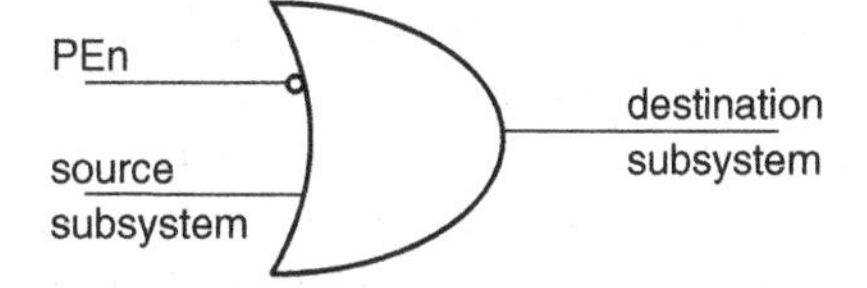

Fig. 5.15b Isolation cell to send a "*1*" during power-off

Figure 5.15c shows an Isolation Cell, which transmits the normal signal when Powered, and, on Power Down, it retains the immediately preceeding value.

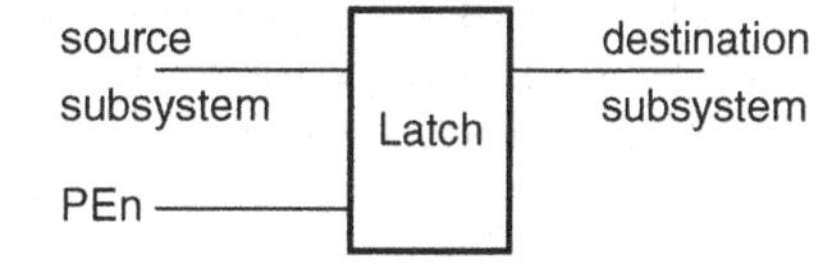

Fig. 5.15c Isolation cell to retain the last value

Figure 5.16 now shows a complete hookup of the interaction among the 2 sub-systems.

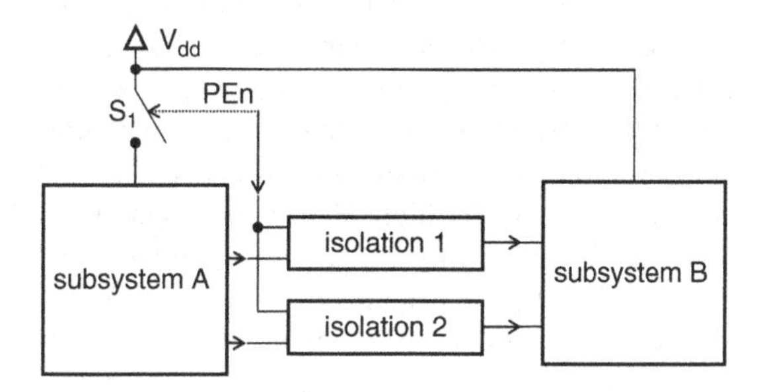

Fig. 5.16 Hookup of the isolation cells

It is possible to use different types of Isolation Cells for the different signals – depending on what is expected by the corresponding input at subsystem B.

5.6.2 Generation of Power-Enable

Once a block of logic is powered off, it would need to be turned back on. Imagine a situation, where the logic controlling the power-enable itself is powered off. The PowerEnable can now never be turned back ON. That means that this logic will never be able to turn ON – once it has been turned OFF. Typically, parts of a circuit whose power supply don't have the capability of switching are called *Always-On*. It is very important that the logic generating the PowerEnable signal should be Always-On.

5.6.3 *Power Sequencing*

The power up and power down can not be done in any arbitrarily manner. Either of these activities have to be done in a specific sequence. For Power Down, the isolation has to be activated in the first step. The driver side can be turned off, only after the isolation has been activated. Similarly, during Power Up, if a very high number of devices were turned off, they can not be turned on simultaneously. A sudden turning on of too many devices will cause a high current to rush to this part of the device. This sudden surge in current is called *rush current*. This rush current causes a voltage drop at other portions of the circuit. Hence, the turning on is done sequentially, with a small portion being turned on at a time. And, isolation is turned off – after the entire portion has been turned on. Sometimes, sequencing might also be required based on the functionality of the circuits and their interdependence on them. Thus, for power up or power down a specific sequence needs to be followed. This is called *power sequencing*.

5.7 Load Capacitance

Switching Power is directly proportional to the load capacitance. There are some gates like *AND, NAND, OR, NOR, XOR, XOR* etc. where the inputs can be interchanged without any change in functionality. So, among the two signals of an AND gate, if one is expected to switch more often than the other, then, the signal with higher activity can be connected to the pin with lower pin-capacitance. This method is not very popular. Usually, the capacitance differential between the two inputs of such gates is not very high. Hence, the saving is not very high. Besides, such decisions can be taken only at the gate-level netlist, when the choice of specific gates and their exact pin capacitances are known. Also, as explained in Section 3.3.1.2, sometimes the critical paths might be using pins with lower capacitance based on timing considerations. Hence, this approach might not be usable, if the signals being interchanged are in the critical path. Theoretically, this also says that among the various alternative gates available, we should choose the gate with lower capacitance. Usually, this would anyways be the case, because of timing considerations. Looking closely, clock-gating also achieves the same effect. For high-activity signals (like clocks), instead of it seeing the loads for all the flops' clock terminals, it sees the load for just one Integrated Clock Gating cell (where the clock is being stopped).

5.8 Input Transition

The instantaneous short circuit path is established for the duration that the signal at the CMOS gate is transitioning. Hence, in order to reduce the duration of such instantaneous short circuit power, the transition should be very sharp. This in turn means, higher drive strength of the previous stage. This in turn means higher current through the previous stage. So, it is difficult to say, whether the net effect will be an improvement or not. Obviously, this adjustment or tuning of Input Transition rate for power consideration is not practiced.

5.9 Device Characteristics

Transistors with higher threshold voltage (also called – *High* V_t) have lower leakage currents. Hence, they are more apt for reducing the leakage currents. However, these have lower speed. So, these can be used only on paths which are not critical. Usually, any adjustment in the device characteristics that reduces (improves) power also reduces (deteriorates) speed.

Usually, combinations of cells are used – Higher performance (and higher power) devices along the critical paths, and, lower performance (and lower power) devices along the non-critical paths. The final choice of a specific device would depend on Power and Speed requirements – and also considering the activity levels of the portion of the circuit. For example if a portion of a circuit has a higher activity, the leakage component is quite low, compared to the active power. In such a situation, instead of choosing a high V_t, it would be better to choose cells with lower drive strength (implying lower short-circuit power). On the other hand, if the portion has very low activity, then, the leakage component could be dominant. In that case, a higher V_t device should be preferred (assuming, there is enough slack on timing).

5.10 Power Estimation

Power Estimation refers to estimating or computing the amount of power consumption. Depending upon the level of accuracy desired, it can be done at various stages of the design. The later you do the estimation in the design cycle, it is expected to be more accurate. However, this more accurate analysis takes that much longer time. More importantly, the further down you are in the design process, your options to take corrective action start reducing. First of all, changes at this stage become more difficult. Worse, the impact of any change made at a late stage is very low. Figure 5.17 shows the same concept through a qualitative graph. For example, an estimate of power at Architecture level could be relatively gross, but, you could take decisions related to voltage or frequency scaling, which are going to have major impacts. On the other hand, at transistor level, though the estimate is much more accurate, there is very little that you can do, if you want the circuit to have lesser power.

Generally, very accurate power numbers at transistor levels are not computed. This is one of the reasons, why, it is called, "estimate". There is one more reason – which is explained towards the end of Section 5.12. Power Estimation at gate level is a relatively mature technology. Power Estimation at gate-level is relatively simple

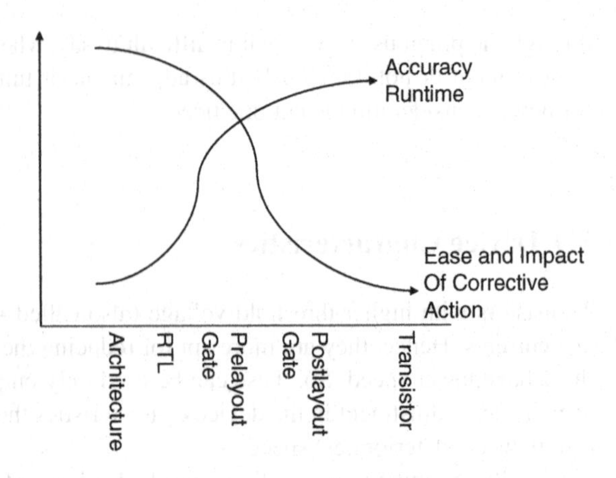

Fig. 5.17 Power estimation at different stages of design

once an STA has been done. The activity at each node is known from annotation of simulation data.

A cell's characterized library data contains several look up tables. The actual table data contains:

- Internal power consumption for each input pin's transitions. Usually, this table is single dimensional, depending only on input transition rate.
- Internal power consumption for each output pin's transitions. Usually, there are separate tables corresponding to each input pin which causes the transition at the output pin. These tables would be 2-dimensional; the two dimensions being output load and input transition time.
- Leakage power for a cell. Usually, this would be just a single entry (rather than a table). However, there would be distinct entries for various combinations of cell-state.

For example, If we consider a simple AND gate, the power-tables would be:

- A single dimensional table for input pin A (say: Table A).
- A single dimensional table for input pin B (say: Table B).

The excerpt below shows an example of a single dimensional table for an input pin in a hypothetical power library:

```
pin(A) {
  direction : input ;
  ....
  internal_power() {
   fall_power(power_template_5) {
     index_1 ("0.01, 0.2, 0.40, 1.20, 2.20") ;
     values("0.06, 0.07, 0.08, 0.10, 0.20") ;
   }
```

```
        rise_power(power_template_5) {
         index_1 ("0.01, 0.2, 0.40, 1.20, 2.20") ;
         values("0.06, 0.065, 0.075, 0.09, 0.18") ;
        }
    }
```

- A two-dimensional table for output pin Y when the transition on Y is due to a transition on A (say: Table C)
- A two-dimensional table for output pin Y when the transition on Y is due to a transition on B (say: Table D)

The excerpt below shows an example of a two dimensional table for an output pin in a hypothetical power library:

```
pin(Z) {
    direction : output ;
    ...
    internal_power() {
     related_pin : "A" ;
     fall_power(power_template_5_5) {
     index_1 ("0.01, 0.02, 0.04, 0.08, 0.16") ;
     index_2 ("0.01, 0.20, 0.40, 1.20, 2.20") ;
     values( "0.17, 0.18, 0.19, 0.23, 0.31",\
             "0.18, 0.19, 0.20, 0.24, 0.32",\
             "0.19, 0.20, 0.22, 0.25, 0.33",\
             "0.20, 0.22, 0.24, 0.26, 0.34",\
             "0.22, 0.25, 0.27, 0.28, 0.35") ;
     }
     rise_power(power_template_5_5) {
     ....
      }
    }
}
```

- A leakage value when the cell output is 0 (say: Value E)
- A leakage value when the cell output is 1 (say: Value F)

The excerpt below shows an example of the leakge specification in a hypothetical power library:

```
cell (myAND) {
.....
leakage_power(){
   value : 5.1e-06;
   when : "A*B";
  }
```

```
leakage_power(){
  value : 6.1e-06;
  when : "!(A*B)";
 }
}
```

5.10.1 Internal Power Estimation

Once the STA is done, the slew values at all inputs of all the cells are known.
So, for each cell:

For each input:
Multiply the number of transitions for that input pin by corresponding power number from the table.
This gives the internal power for the cell – contributed by the transitions at this input pin.
Add this pin's contribution to the value obtained so far.
End For

For example, for an *AND* gate, say, pin *A* has *T1* transitions and pin *B* has *T2* transitions.
So, internal power contributed by $A = T1$ * entry from *Table A*.
Internal power contributed by $B = T2$ * *entry from Table B.*
Sum of these two values gives the internal power consumption – due to inputs transitioning.

The effective load capacitance seen by a cell output can easily be known through STA.

For each output:
Understand which transition on an output was caused by which input. The establishment of this correlation is not deterministic. Each power analysis tool uses its own mechanism to establish this relation.
For each input causing a transition at this output
*Multiply the number of transitions_caused_by_input_pin * corresponding power number from the table for the output pin. This gives the internal power of the cell – contributed by the transition on the input-output combination.*
End For. This gives the internal power contribution due to the specific output pin
End For. This gives the internal power contribution due to all output pins of the cell

Say, pin *Y* has *T3* transitions. Out of these *T3* transitions, *T4* have been triggered due to a change in *A*, and, *T5* have been triggered due to a change in *B*.

$$T4 + T5 = T3.$$

Internal power contributed by *A->Y* transition = *T4* * entry from *Table C*.
Internal power contributed by *B->Y* transition = *T5* * entry from *Table D*.

Sum up all the numbers obtained so far. This is actually the energy consumed. Divide this by the duration for which the transitions were measured.

This gives the Internal Power.

5.10.2 Switching Power Estimation

For each output, STA analysis already computes the effective capacitance for that output. The activity data tells the number of times the output switched. Compute: $\frac{1}{2}CV^2$*Number_Of_Transitions. This gives the total Switching Energy. Divide this by the duration for which the transitions were measured. This gives the Switching Power. Look a bit more closely at $\frac{1}{2}CV^2$ per transition. During High to Low transition, this is the amount of charge stored in the capacitor that got dissipated in the N-MOS. Also, during Low to High transitions, when the capacitor was getting charged to store $\frac{1}{2}CV^2$, an equal amount of energy got dissipated at the resistive component of the P-MOS, through which the capacitor was getting charged. Thus, over two transitions (one charging and discharging), a total of CV^2 got dissipated, which can be thought of as $\frac{1}{2}CV^2$ per transition.

5.10.3 Leakage Power Estimation

From the simulation data, it is known for how long the circuit stayed in which state. The Leakage Power for that state is known from the cell's library data. This leakage Power is aggregated to get the Leakage Power numbers.

5.10.4 Power Estimation at Non-gate Level

At the transistor level, a more accurate Power Analysis is possible, through transistor level simulations. However, these take too long to run. Usually, a need for full-blown power estimation at the transistor level is rare. Only some specific aspects might be analyzed at transistor level, if there is a need.

Since late 90s, attempts have been made to Estimate Power at RTL level also. Today, there are some tools which are able to do a fairly accurate estimate of Power at the RTL-levels also. Over the last decade or so, these tools have become fairly sophisticated in the sense that they make use of technology library to get the cell power for the specific technology nodes, and they also use timing constraints, so that they can do a realistic choice among low-drive and high-drive cells.

Several design houses have spreadsheet based utilities to do a very rough estimation of power. Also, there are some commercially available tools to do power estimation at architectural level. These are still fairly inaccurate. However, at this level, the runtimes are very small. As explained towards the beginning of the section, the earlier we do the estimate, the lesser would be the accuracy. Smaller runtimes allows you to explore various possibilities. Thus, an analysis earlier in the design flow is used very frequently – mostly to do comparative study among various alternatives. At this stage, the relative values (or, savings) are of more importance than the absolute power numbers.

5.11 Probabilistic Estimation

Probabilistic Estimation refers to a situation when the Activity data is not annotated from simulation, rather the activity is computed probabilistically. All other computations remain the same as explained in Section 5.10.

Let us use the notation:

$S(A)$ = Number of transitions at A (within a given time period)
$P(A)$ = Probability that A is at value 1

Consider an *AND* gate, with inputs A and B; and output Y. Y will be at *1*, when both A and B are at *1*. Hence, $P(Y) = P(A)*P(B)$ Any transition at A will go to the output, only if B is *1*. So, $S(Y)$ *(due to change in A)* $= S(A)*P(B)$. Similarly: $S(Y)$ (due to a change in B) $= S(B)*P(A)$. We can do a sum of both these to get the final activity at Y.

Consider the following example of transitions for the AND gate:

A	B	Y
0	0	0
0	1	0
1	0	0
1	1	1

From the transitions and state mentioned in the table,

$P(A) = 0.5$; $P(B) = 0.5$
$S(A) = 1$; $S(B) = 3$

Now, $P(Y) = P(A)*P(B) = 0.25$. This matches the table entries for Y. Y is at *1* for only 25% of the time.

$S(Y)$ *(due to change in A)* $= S(A)*P(B) = 0.5$
$S(Y)$ *(due to change in B)* $= S(B)*P(A) = 1.5$

So, $S(Y) = 2$, which is different from the actual number of transitions seen on Y – in the table. This difference can be explained by the fact that when A and B both simultaneously changed from 01 to 10, Y might have had a glitch!!!

Similarly, for an *OR* gate:

$$P(Y) = P(A) + P(B) - P(A)^* P(B)$$

because, *Y* is *1* when either of *A* or *B* is *1*. The last term is to cancel out the double counting, when *A* and *B* are both simultaneously at 1.

$S(Y)$ (*due to change in A*) $= S(A)^* [1 - P(B)]$; a transition on *A* will reach *Y* only if *B* is not at *1*.
$S(Y)$ (*due to change in B*) $= S(B)^* [1 - P(A)]$; a transition on *B* will reach *Y* only if *A* is not at *1*.
$S(Y) =$ Sum of the values computed above.

Compute also the probabilistic propagation through an XOR gate.

$P(Y) = P(A)^* [1 - P(B)] + P(B)^* [1 - P(A)]$
Y will be *1*, when (*A* is *1* and *B* is *0*) or (*B* is *1* and *A* is *0*)
$S(Y) = S(A) + S(B)$
Any transition on *A* or on *B* will reach *Y*

Similarly, it should be possible to compute the Switching estimates at the outputs of various gates, if we know the:

- Switching activities at its various inputs
- Probability of each input being at 1

Actually, for Power Estimation, only switching activity is sufficient. However, Probability of *1* is required in order to compute the switching activity at the next stage.

5.11.1 Spatial Correlation

Probabilistic Estimates could go wrong in case signals are spatially correlated. If there are two signals which have some relationship amongst them, the Probabilistic Estimate might not be able to keep track of that relationship. Consider the example of a simple MUX for the computation of Probability only. You have already seen that if the probability computation is wrong, the switching activity computation at the next stage goes wrong. Assume all the inputs can take each possible combination with equal probability. So, the possible truth table would be:

A	B	S	Y
0	0	0	0
0	0	1	0
0	1	0	0
0	1	1	1
1	0	0	1
1	0	1	0
1	1	0	1
1	1	1	1

$$\text{So, } P(A) = P(B) = P(S) = 0.5$$

Also, you can see that for the output $P\,(Y) = 0.5$
Consider an implementation of the MUX as shown in Fig. 5.18.

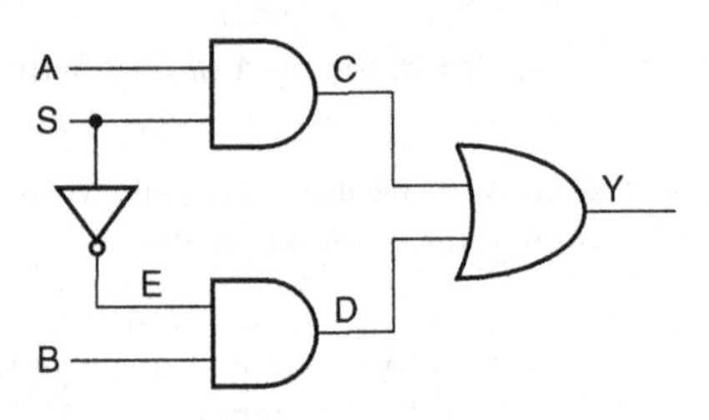

Fig. 5.18 Mux implementation

We will compute *P(Y)*.
$P\,(C) = P\,(A)^*\,P(S) = 0.25$
$P\,(E) = 1 - P(S) = 0.5$
$P\,(D) = P\,(E)^*\,P(B) = 0.25$
$P(Y) = P(C) + P(D) - P(C)^*P(D) = 0.5 - 0.0625 \mathrel{!=} 0.5$

Thus, the computed value has come out to be incorrect!!!
The reason for the incorrectness is:

C and *D* can never be *1* simultaneously. This is a spatial relationship between *C* and *D*. While, computing *P(Y)* a subtraction factor has been applied to account for the situation when both *C* and *D* are simultaneously *1*. As per the spatial relation among *C* and *D*, this factor has to be *0*. But, since this spatial relation was not kept track of, a correction factor of 0.0625 got applied – resulting in incorrect value.

5.11.2 Temporal Correlation

Temporal Correlation refers to a situation, where, signals follow certain specific pattern in time. However, this cyclic pattern is not known while doing the probabilistic computation. Consider a signal which goes High once in 6 cycles. So, its probability of being at *1* is *1/6*. Figure 5.19 shows the circuit for the realization of such functionality.

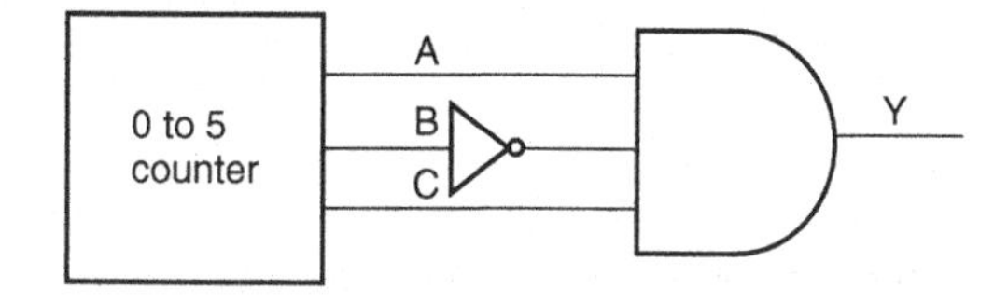

Fig. 5.19 *Y* goes "High" once in 6 cycles

The truth table for one complete cycle of the counter would be:

A	*B*	*C*	*Y*
0	*0*	*0*	*0*
0	*0*	*1*	*0*
0	*1*	*0*	*0*
0	*1*	*1*	*0*
1	*0*	*0*	*0*
1	*0*	*1*	*1*

So, $P(A) = 2/6 = 1/3$
$P(B) = 2/6 = 1/3$
$P(C) = 3/6 = 1/2$
$P(\text{at the output of inverter}) = 1 - P(B) = 2/3$
$P(Y) = P(A)^{*}P(\text{at the output of inverter})^{*}P(C) = 1/9 \mathrel{!}= 1/6!!!$

Again, because the cyclic nature and the exact pattern is not kept track of, hence, the computation at *Y* becomes incorrect.

5.12 Simulation Plus Probabilistic

More often than not, Power Estimation at the gate level is done using a combination of simulation and Probabilistic. Say, a design is simulated at RTL. The activity data is captured based on these simulations. The design is now synthesized. Due to synthesis, there are a lot of additional nodes being created. However, a lot of nodes present in the RTL are also present in the gate-level netlist. These nodes are Registers, Memories, ports etc. The Power Estimator tool reads in the simulation data (obtained from RTL simulation). For the nodes which are also present in the netlist, the activity information is applied based on the simulation data just read. For the remaining nodes in the netlist, the activity is computed based on probabilistic propagation.

The advantage of this mixed approach is that, even if some error is introduced due to probabilistic propagation, this error will not propagate too far. Very soon, the estimation engine will encounter a node that is annotated with the activity data. So, any computed error will not propagate beyond this point. In Fig. 5.20, the activity at ports *A, B, C, D, E* and at the output of registers *F1, F2, F3, F4* and *F5* will be annotated based on simulation data (obtained from RTL simulation).

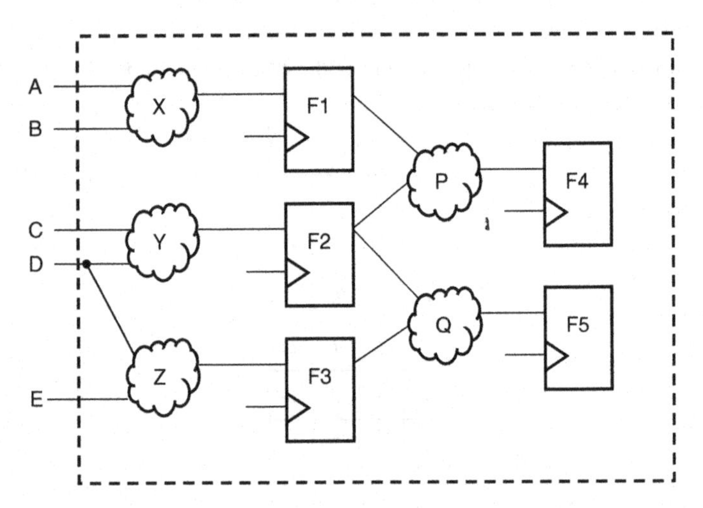

Fig. 5.20 Mix of probabilistic and simulation based analysis

The activities within clouds *X, Y, Z, P* and *Q* would be computed based on probabilistic propagation. Even if an error is introduced in the propagation within *X*, it will not propagate beyond *F1*, because, at *F1* the value is annotated directly from simulation data. This mix of Probabilistic and Simulation based methodology allows Power Analysis to be performed with reasonably high level of accuracy, even in the absence of gate-level simulation data.

By now, it should be fairly apparent to you – that the power value estimated is dependent significantly on the vector-set applied. Thus, the vector set used for power estimation should be chosen – such that it mimics the activity of the actual application. It should be understood that it is not a good idea to use the same vector set that was developed for functional simulation based verification. For the purpose of simulation based verification, the vector set is developed with the aim of exhaustive coverage – including corner cases. A corner case situation as well as the situation of general use has equal importance for functional verification. However, for power estimation – the vectors should be so developed – that they keep the device for longer duration in the usage mode in which the device is generally expected to perform.

Thus, any power number computed – is true only for the specific vector set. A different usage pattern, and, the power number would change. This is one of the reasons, why the whole analysis is just an estimate. Another reason was given towards the beginning of Section 5.10.

5.13 CPF/UPF

CPF stands for *Common Power Format*, and is proposed by SI2. UPF stands for *Unified Power Format*, and is proposed by Accellera.

There are certain intents and information related to Power aspects which are not specified in RTL. CPF and UPF are formats for specifying such intents and information. Examples of information that can be conveyed through these formats include:

- Portions of design which operate at different voltage levels
- Identification of level shifters and their respective voltage levels
- Identification of isolation cells and their nature (transmit 0, transmit 1, retain last value etc.)
- Functionality expected from an Isolation Cell on a specific module boundary, when, it has been powered down
- Switches for Power Down etc.

This information is used by:

- Synthesis Tools
- Formal Tools
- Rule Checkers
- Simulators
- Layout Tools

Examples of usage of such information include:

- Synthesis tools need to know the voltage levels, so that it can compute the delays etc. correctly
- Formal tools and rule checkers need to validate that the design realized matches the intent of the designer. In addition, they would need to validate that some fundamental principles are not violated, e.g. PEn signals should only be generated from AlwaysON portions of design (i.e. never Powers Down).
- Simulators can use the information to "model" certain behavior related to Power Supply. Example, when the PEn signal goes Low (means, a block is no longer

receiving power), a simulator can corrupt all the outputs of this block. This can help in validating that the isolation cells are connected properly, and are ensuring correct operation of the downstream circuit – even when a specific portion has been shut-off.

- Layout tools can use the information to connect the correct power rails.

Chapter 6
Design for Test (DFT)

After the design is fully validated for functionality, timing etc. through a lot of rigorous analysis, it is sent for fabrication. After fabrication, the fabricated part is subjected to further testing.

6.1 Introduction

Design For Test means keeping testing related aspects under consideration during the design stage itself. In the context of *DFT*, test means testing for manufacturing defects. As far as testing the design for desired functionality is concerned, that is anyways validated through simulation, assertion, rule checking based techniques etc.

6.1.1 Manufacturing Defect – Vis-a-Vis – Design Defect

Suppose, you want to realize the functionality of a *counter*. You use techniques like simulation, assertion etc. to ensure that your design really functions as a counter. However, once the design is put on silicon (fabrication), you want to ensure that even though the design was supposed to be working as a counter, the fabricated design actually does behave as a counter. This is because, due to some manufacturing defects, some of the devices might not really function as designed. The manufacturing defects stem mainly because many of the processes involved in the fabrication need a very fine precision, and, due to statistical variations, sometimes, there might be defects introduced in some of the devices. These defects could either cause two lines in close vicinity to be *shorted*, or, a line to be broken (resulting in an *open*); or sometimes, a specific junction might not have been fabricated correctly.

One way to do the testing is to check the whole fabricated device and ensuring that it is really behaving correctly as a *counter*. Another way is to check whether each of the constituent elements really acts as what it was expected it to be. For manufacturability fault tests, the second approach is used, i.e. instead of validating whether the complete circuit acts as a *counter* or not, you validate for each individual

S. Churiwala, S. Garg, *Principles of VLSI RTL Design*,
DOI 10.1007/978-1-4419-9296-3_6, © Springer Science+Business Media, LLC 2011

gate, whether the *AND* gate really is behaving like an *AND* gate; whether an *OR* gate is really behaving like an *OR* gate, etc. If each individual component is found to be behaving correctly, it is expected that the whole device will actually work properly in terms of its overall functionality – because of all the verification that has already been done – for the design itself.

6.1.2 Stuck-At Fault Models

Manufacturing defects could be of several types. In this book, we will consider mostly the *stuck-at fault* model. Once this model is well understood, you should be easily able to appreciate the other fault models, used against detection of various manufacturing defects. *Stuck-At fault* model assumes that any given net (or, pins of various component elements) could be stuck – either at a *0* (i.e. *shorted* with *Ground*), or, a *1* (i.e. *shorted* with V_{dd}). Hence, testing against this fault model effectively means testing each net (or, pins of various components) to validate that it is free to move to the value as desired by the circuit functionality, rather than tied to either a *0* or a *1*.

6.1.3 Cost Considerations

Usually, validation of functionality (as you understand from simulation etc.) is done on a per design basis. Once a design is known to be verified for correct functionality, the same verification is not required for each manufactured part. But, the validation against manufacturing defects has to be carried on for each manufactured part. For each of these manufactured parts, you need to validate each pin of each included gate. For each pin, you need to validate that it is free to move to *0* (i.e. not stuck-at *1*) and also to *1* (i.e. not stuck-at 0). So, there is a lot of testing to be done. Because, this test against manufacturing defect has to be done on a per-part basis, hence, the time and the cost to test each part is a very significant consideration. The devices used for testing the manufactured parts called *testers* are also very costly. That is another motivation for the parts to be tested in as little time as possible – so that costly tester machines are not tied too long on each of the devices. This puts a set of requirements on you – as the RTL designer. Hence the term, Design For Test (DFT). If you are careful of these requirements, the testing can be easy (which means, lesser time on tester) – thereby, impacting the overall cost of the manufactured chips.

6.2 Controllability and Observability

The testing for each pin against stuck-at situations is conceptually very simple. Consider an *AND* gate shown in Fig. 6.1.

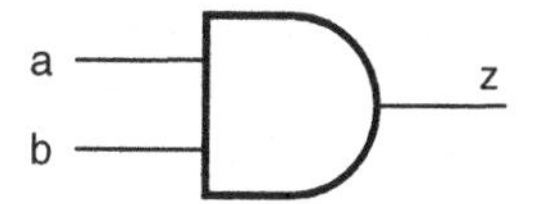

Fig. 6.1 Stuck-at fault testing of an AND gate

You want to check whether its output is stuck-at *1* or not. So, you apply a *0* at any of its two input pins. For the other input pin, you can apply whatever value you want. Now, you check the output of the *AND* gate. If the output is *0* then you know that this output pin is not stuck-at *1* (because, it can go to *0*). Similarly, for testing the same output to be not stuck-at *0*, you apply a *1* at both its input pins. If the output is found to be at *1*, you know that the output is not stuck at *0*. Now, you want to check for similar stuck-at situation on one of its input pins (say *a*). You will apply a *1* at the other input (*b*). Now, you will put a *1* on *a*. If the output *z* also goes to *1*, you know that *a* really had gone to *1* – which means, *a* was not stuck-at *0*. Next you apply a *0* on *a*. Now, if the output *z* also goes to *0*, you know that *a* really had gone to *0* – which means, *a* was not stuck-at *1*. Similarly, you can check for *b* against such stuck-at faults.

So, for testing the output of the *AND* gate, you need an ability to

- Control any of the inputs to a *0*
- Control both the inputs to *1*
- Observe the value at the output of the *AND* gate

And to test one of the inputs (say *a*) of the *AND* gate, you need an ability to

- Control the other input (i.e. *b*) to *1*
- Control this input (i.e. *a*) to both *0* and *1*
- Observe the value at the output of the *AND* gate

The ease with which you can control the values at the inputs of the *AND* gate and observe the value at the output of the *AND* gate effectively determines how easy it is to test this *AND* gate. So, effectively, the only thing that you (as an RTL designer) need to ensure is that all the inputs to each of the gates are easily controllable, and, the outputs from each of the gates are easily observable. If these two conditions are met, the DFT engineers will be able to generate a set of patterns, and, using those patterns, the tester machine can easily test each gate and thus, validate the entire part against any manufacturing defect. As can be easily imagined, for a simple *AND* gate – meeting the above requirements is fairly simple!!!

6.2.1 Controllability and Observability Conflict

Look at this same AND gate in the context of two more gates in its immediate vicinity, as shown in Fig. 6.2.

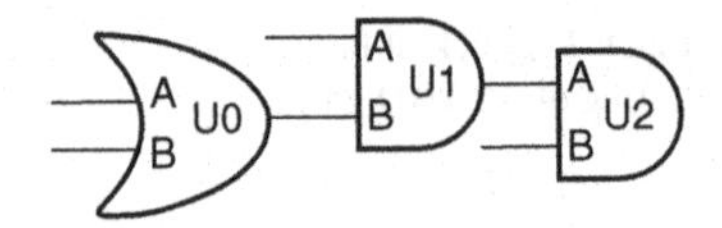

Fig. 6.2 AND gate in the context of its neighbors

U1 is the *AND* gate that you want to test for manufacturing defects. Now, in order to control *U1/B* to *0*, you need to control both inputs of *U0* to *0*. Similarly, in order to observe the value of *U1*, you need to control *U2/B* to a *1* – so that *U1's* output can pass through *U2*. On the other hand, if you have to test *U0* for manufacturing defects, it is very easy to control *U0's* input pins. But, in order to observe *U0's* output, you have to ensure *U1/A* is at *1* and *U2/B* is also at *1*. And, if you have to test *U2*, observing its output is fairly simple. But, controlling its inputs might need controlling values at *U1/A*, *U0/A* and *U0/B*.

Effectively, the nearer a gate is towards the input, the easier it is to control. However, the nearer it is towards input, the farther it becomes from the output, and hence, becomes that much more difficult to observe. Similarly, the nearer a gate is towards the output, the easier it is to observe. But, that makes it farther from the input – making it more difficult to be controlled. So, for any gate with better controllability (meaning, proximity to input), the observability becomes bad; and for any gate with better observability (meaning, proximity to output), the controllability becomes bad.

6.3 Scan Chains

Scan Chains refer to a set of flops or latches connected together in the form of a chain.

6.3.1 Need for Simultaneous Control

In a real design, there are millions of gates and many of them are even sequential devices (i.e. flops). Contrary to the simple diagrams that you have been seeing all through the book, the signals just don't move in one direction from one flop to the next one and so on. They crisscross all over the place, including going through many sequential devices. Figure 6.3 shows a hypothetical diagram. Stars in this diagram represent flops. Dots represent combinational gates.

Gate *Q* is easy to control. But, if its values have to be taken to an output (so that it can be observed), it will require a lot of other gates in the path (till the output) to be made transparent (by controlling the other pins of all those gates). *P* is easy to observe. But, if its inputs need to be controlled at specific values, it will require a lot of gates in the path (from the input) to be made transparent. For *R*, both controllability and observability requires many other gates to be made transparent.

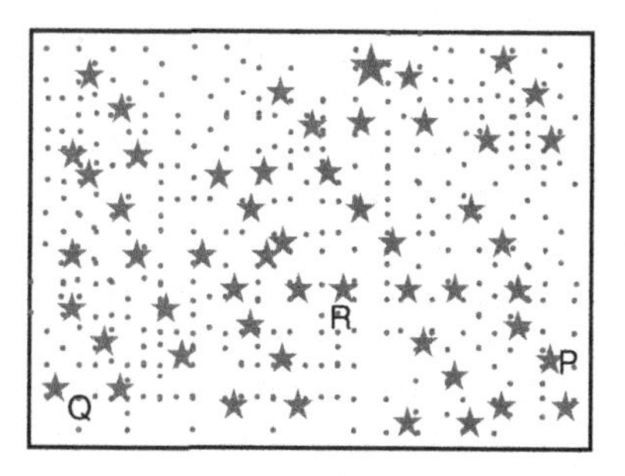

Fig. 6.3 Complex chip having many gates

6.3.2 Complexity Due to Signal Relationship

Many signals in the design have relation in space. That means, if a signal is at a given value, it will also imply some other signal to be at a given value. Consider the circuit shown in Fig. 6.4a.

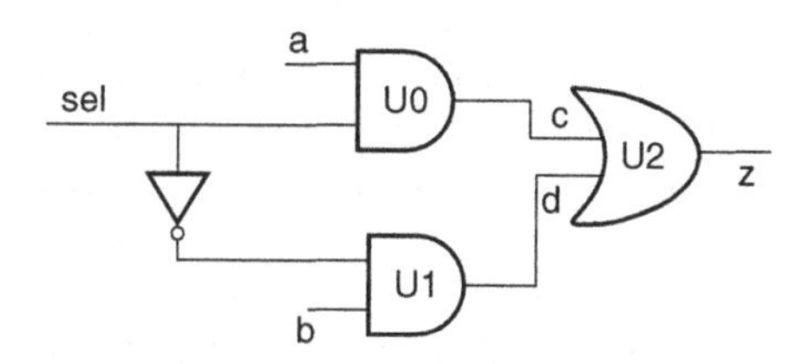

Fig. 6.4a Spatial relationship

A value of *1* at *c* requires that *sel* is also *1*. So, the output of the invertor is *0*. That means, *d* is also at *0*. Thus, a *1* at *c* imposes that *d* is at *0*. This is an example of a spatial relationship. Such spatial relationships mean – it is not always possible to control the inputs of each gate independently. Putting a value at some input may cause some other input of the same or a different gate to become uncontrollable. For example, in Fig. 6.4a, if you control *U2/A* to a *1*, it automatically means *U2/B* is uncontrollable (it will be forced to a *0*).

Similarly, temporal correlation also adds to the challenge. In Fig. 6.4b, a *0* at *U0/A* also ensures a *0* at *U1/A* in the immediately following cycle.

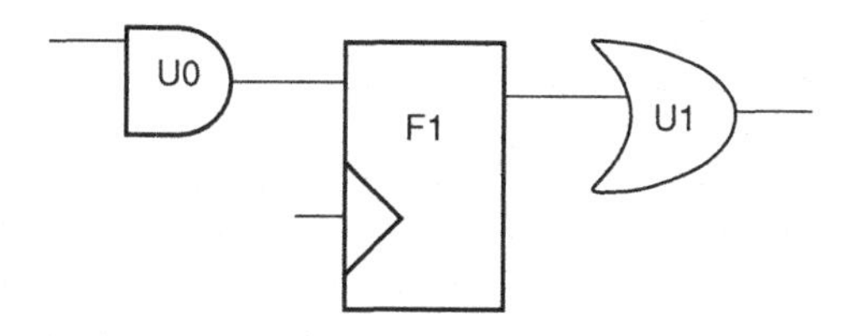

Fig. 6.4b Temporal relationship

So, by taking specific signals to some values, some other signals could become uncontrollable either in that cycle itself, or, in another cycle. Thus, the more is the number of pins to be controlled simultaneously, the more difficult it becomes.

6.3.3 Need for Many Control and Observe Points

Obviously, one simple solution to all these complexities around controllability and observability is: you should have plenty of inputs and outputs. More inputs can reach more places within the circuit, including some of the nets which are too deep. Similarly, more outputs are able to tap more places within the circuit, including some of the nets which are too deep.

Refer again to Fig. 6.3. If you had plenty of inputs, then, in order to control *P* or *R*, you could use one of the inputs which feeds very close to *P* or *R*, and then the value has to be moved across just a few gates – so that it can reach *P* or *R*. And, for making those few gates also transparent, you can control them relatively easily, by directly putting in the values – very close to them – through these inputs – which are now all over the design.

Similarly, if you had plenty of outputs, then, in order to observe *Q* or *R*, you could move the value across just a few gates and you would reach a point which can be directly tapped to one of the outputs. Thus, the problem of observability and controllability would be reduced significantly. However, adding so many inputs and outputs is not so easy. Chips are already short of space to put in just about enough pins required for their functionality. So, there is no space on their periphery to put in so many more pins. Besides an increase in die-size (to accommodate so many more pins on the periphery) there will also be an increase in package costs.

6.3.4 Using Scan Chain for Controllability and Observability

Since, it is difficult to put in so many extra inputs and outputs for better observability and controllability, thus, as an alternate mechanism, all the flops in the design are connected in series. The first flop in the series is fed directly from an input pin. And, the last flop in the series feeds directly to an output pin. These flops in the series form a *scan chain*. Figure 6.5 shows a simplified excerpt from a design. This is the design required in order to achieve the desired functionality.

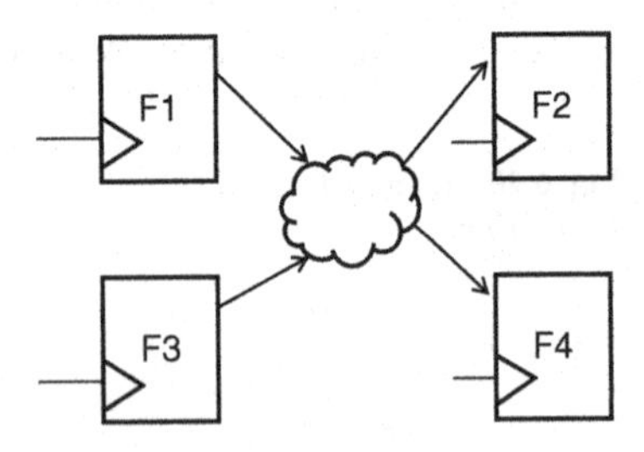

Fig. 6.5 Normal functionality

Figure 6.6 shows the same design with the scan chain inserted. The dotted lines represent the scan paths. A scan path has been established along *SI* (primary input) *F1 F3 F4 F2 SO* (primary output).

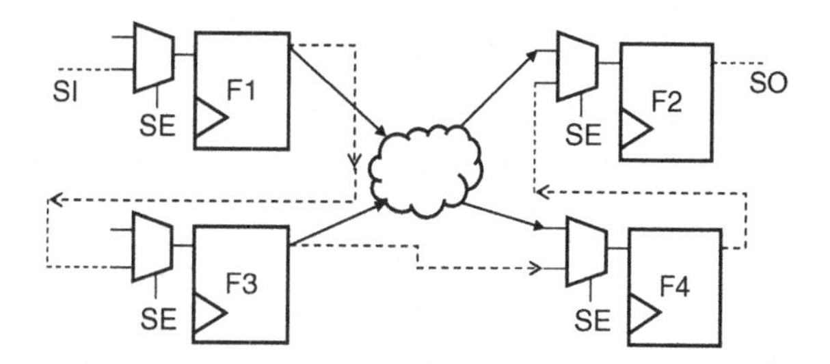

Fig. 6.6 Scan chain

Now, if you have to put a specific value at *F2*, it is relatively simple. The value can be put onto *SI*, and, 4 clocks applied. The value would have reached *F2*. Similarly, if the value at *F1* has to be observed, simply apply 3 clocks, and, the value will be available on *SO*. Signal *SE* which is common to all the flops in the chain controls the *mux* to decide whether the path along the scan chain will be activated or the path along the regular functional operation would be activated. So, each of these flops in the chain is acting almost as a Primary Input, because, it can be assigned any given value – irrespective of whatever else is there on any other flop. These flops are all over the design. So, there is a huge increase in the controllability of the design. A flop which is deep down in the chain will need more cycles to get its value, but, it can get the value without any interference from any other device.

Similarly, each of these flops in the chain is also acting as a Primary Output, because, its value can be observed – irrespective of whatever else is there on any other flop. These flops being spread all over the design, there is a huge increase in the observability of the design. A flop which is further up (nearer to SI) in the chain will need more cycles – for its value to reach the output, but, its value can be observed without any interference from any other device. Thus, the impact of creating the scan chain is that it has now become much easier to both control as well as observe any point in the design, irrespective of where the point lies.

6.4 Mechanics of Scan Chain

The process of creating the scan chain can be fairly automated. The principles mentioned in this chapter are to ensure that the chains can be created almost mechanically.

6.4.1 Scan Flop

In Fig. 6.6, you created additional paths – which form the scan chain – by inserting muxes in front of all the flops. In reality, the mux is included inside the library cell. Such flops with the scan mux included are called *scan flops*. Figure 6.7 shows the symbol for a regular flop (on the left) and the corresponding scan version (on the

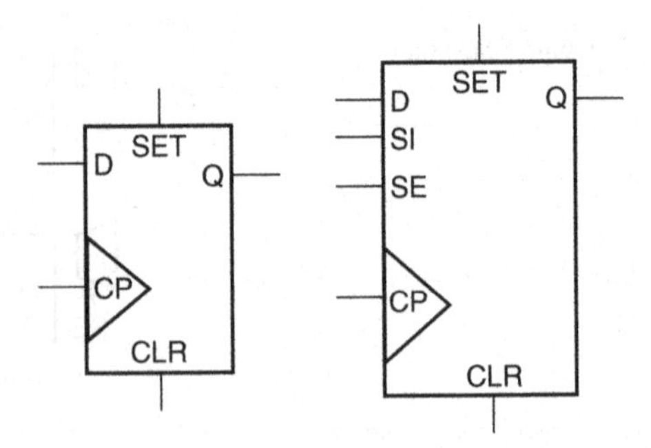

Fig. 6.7 Normal and scan flops

right). If *SE* is active, the value at *SI* (needed for DFT) will get captured inside the flop, and, if *SE* is inactive, the value at *D (functional value)* will get captured inside the flop.

6.4.2 Replacing Flops with Scan Versions

First the synthesis is done for normal functionality, without any consideration for scan paths. At this stage, synthesis tools have to be specifically directed not to use scan flops. The functionality of a scan flop is equivalent to a mux plus flop. If at this stage, scan flops are not barred from taking part in realizing the functionality, it is possible that the synthesis tool might make use of a scan flop – if it has to infer a mux just before a flop. Once the normal functionality has been achieved, all the flops have to be mechanically replaced by the corresponding scan version from the library. This can be done using scripts. This step is called *scan insertion*, and, is one of the easiest steps in the entire process related to scan. *Scan insertion* is followed by *scan stitching* (explained in Section 6.4.5), which means hooking up the scan flops into a set of chains. *Scan insertion* should also explain the reason for disabling use of scan flops during synthesis. If a scan flop has already been used for normal functionality, it can no longer be converted to its scan version, which means, it can not be a part of the scan chain.

6.4.3 Timing Impact of Scan Flops

Scan flops have higher *setup requirement* (even for functional paths), as, the functional data has to go through an additional mux (embedded inside the scan flop). Hence, it is possible that a path which was meeting the timing before scan insertion starts failing after scan insertion. This higher *setup* would be seen – even before the chain itself has been stitched. Just replacing the regular flop with scan flops will show up this effect of higher setup time. In many libraries the scan flop has an additional output (generally referred as SO pin) – so that the regular output Q does not suffer any additional load due to chain stitching. The chain stitching would use SO, while, regular functionality would use Q.

In some libraries, there is no distinct SO pin. Such cells use Q for both regular functionality as well as scan-chain stitching. So, a flop which was driving only the functionality now sees an additional load (scan-in) of the next flop in the chain. Due to this additional load, the delay through the flop will increase, thus increasing the path delays. Since this happens on all the flops, so, all path delays (where, the start point is a flop) would increase. This load would be seen after the chain is stitched. This brings in some uncertainty – in terms of the amount of additional delay that would be seen. During scan insertion, you may decide to connect a flop's output to its own SI – so that the loading impact of SI can be considered during timing-analysis, even though, the actual chain is not yet stitched.

6.4.4 Area Impact of Scan Flops

The scan versions of the flops are larger in area, compared to their non-scan counterpart. This increases the overall silicon area of the design. Though, silicon real-estate is very costly, however, this increase in area is a relatively much smaller price to pay, compared to the huge increase in controllability and observability that such flops bring. Sometimes, you might want to allow the use of scan flops during synthesis stage itself. Use of such scan flops for normal functionality might be done to achieve better area or performance. These flops (where, scan is used for normal functionality) will need to be excluded from the scan chain.

6.4.5 Stitching the Chain

The order in which the flops are stitched in the chain is not very important – at least, during the initial stages of the design. During the initial stages, you may connect the chains in random sequence. Or, you may connect the chains based on alphabetical ordering of the hierarchical instance names of the flops. You can use any criterion – it simply does not matter. However, after cell-placement is done, the scan-chains are re-stitched. This time – on the basis of physical proximity of the scan flops while also ensuring that the flops in the same chain are in the same test clock domain. The physical proximity criterion is considered to save on routing needs and also to save on delays on the scan path.

6.5 Shift and Capture

The concept of scan testing depends fully on a series of shifts and captures.

6.5.1 ShiftIn

Consider a specific cell's specific pin is desired to be taken to a given value. You could determine the fanin cone of this pin to the nearest flops. This cone might have *N* flops. You should be able to determine the values on each of these flops that will result in the desired value at the point of interest. You can count the position of each of these flops in the chain. And, you can put the values into those flops by simply pushing in the values in the right sequence at the *SI* pin, and, giving the clocks, till all the values reach the desired flops. Since you want the values to be moved across the flops along the scan chain, so, *SE* is kept asserted. This operation is called a *ShiftIn*. For example, consider the sample circuit shown in Fig. 6.8.

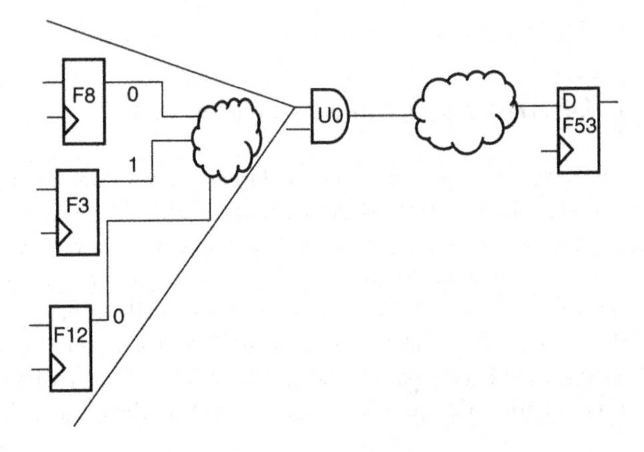

Fig. 6.8 Example scan and capture

Say, you want *U0/A* to get a value of *1*.You traverse the fanin cone of the pin and get three flops. *F<number>* in the schematic denotes the sequence number of the flop in the scan-chain. The value shown at the flop output shows the value that the flop should have, so that *U0/A* gets a *1*. Now, you need to feed the pattern "*0xxx 0xxx x1xx*" into *SI* and apply 12 pulses of clock. The first *0* is for *F12*; the next *0* is for *F8* and the *1* is for *F3*. At the end of the 12 pulses, *F12* has a *0*; *F8* has a *0* and *F3* has a *1*. So, *U0/A* also has a *1*. The space in the pattern is just for ease in readability, and, it does not denote anything. The *x* in the pattern denotes *don't care*. These values would be lying on flops *F1, F2, F4* etc. which do not impact the presence of 1 at U0/A. This completes the *ShiftIn* (or, *ScanIn*). The name *ShiftIn* denotes the input values are being shifted by one flop with each cycle of the clock.

6.5.2 Capture

At this instant, *SE* is de-asserted, and, another clock is applied. This will cause the *U0* output to be sampled by the flop in the immediate fanout. This will be flop *F53* in Fig. 6.8 This is called *Capture*.

6.5.3 ShiftOut

Now, *SE* is asserted again, and, the value stored in the *Capture Flop* is moved along the scan chain till it comes out at *SO*. This is called *ShiftOut* (or, *ScanOut*). By examining this value against the expected value, we can know, if the gate *U0* has some manufacturing defect or not. Figure 6.9 shows a sequence of *shift* and *capture*.

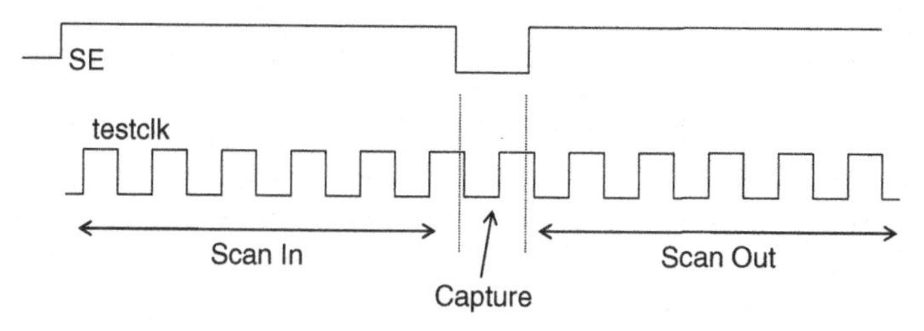

Fig. 6.9 Shift and capture

A mismatch in *ShiftOut* value does not necessarily mean the fault is with *U0*. Rather, the fault could be anywhere in the fanin cone of *U0* (thus, *U0/A* did not get the desired value), or, in *U0* itself (thus, *U0* did not produce the right value at its output), or, in the fanout cone of *U0* (thus, the value produced by *U0* got corrupted before reaching the capture flop). Hence, usually, a series of such cycles are required to pin-point the location of the fault. You have anyways seen (in Section 6.2) that several combinations of values need to be checked to reliably conclude that the gate and its pins are free of manufacturing defect.

6.5.4 Overlapping ShiftIn and ShiftOut

While doing a *ShiftOut*, the *testclk* is anyways running, and, *SE* is asserted. So, this time is also used to *ShiftIn* the new sequence. So, while one *Capture* value is being shifted out, the sequence for the next capture is getting shifted in. Figure 6.10 shows this overlap.

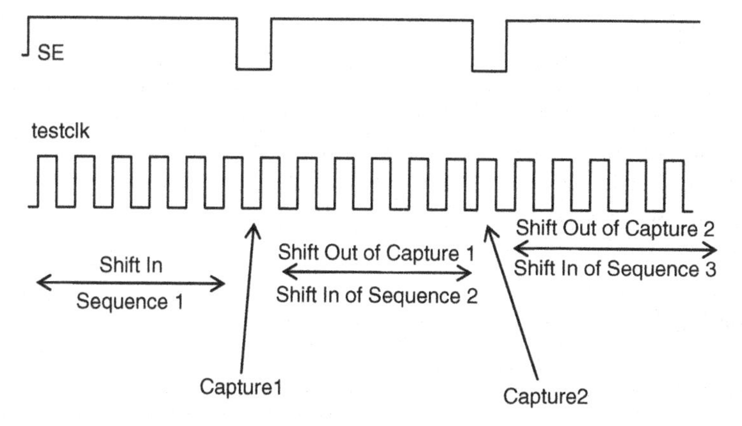

Fig. 6.10 Overlap of *ShiftOut* and *ShiftIn*

Also, you have seen (in Section 6.5.1) that as you want to *ShiftIn* a sequence to put the desired values at a pin, there are a lot of don't cares. These don't cares can be used to put in values on other scan flops (e.g. *F1, F2, F4, F5* etc. in the example of Fig. 6.8), so that they can put in some desired values on some other pins also. This will allow multiple pins to be checked – during a single Capture cycle.

As can be well imagined, obtaining an optimal sequence – so that the Shift cycles are utilized adequately and all the pins are tested against both types of stuck-at situations – using minimal number of patterns is a highly algorithmic problem. There are sophisticated programs called ATPG (Automatic Test Pattern Generation) tools – which create such optimal sequence. As an RTL designer, you do not have to actually generate these patterns. You just have to make sure that your circuit has good controllability and observability – so that the ATPG tools find it easy to generate the patterns. Section 6.6 onwards explains some of the things that you have to take care of in your RTL to ensure good controllability and observability.

6.5.5 Chain Length

A design contains thousands of flops. Imagine all of them being stitched together into one chain. If a value has to be shifted into the last flop in the chain, it is going to take a huge number of cycles. Similarly, if the value has been captured in the first flop in the chain, and, it has to be shifted out, it is going to take a huge number of cycles. This will make the whole process to be very slow. Thus, instead of stitching all the flops into one huge chain, several small chains are created. Usually, a chain has about 125–150 flops. This allows multiple chains to be created. Presence of multiple chains allows multiple *ShiftIn*, *Capture*, *ShiftOut* to be carried out simultaneously. The only requirement is that there has to be multiple *SI* and *SO* pins; one each per chain.

All these multiple chains are also kept to be of similar length. Consider a situation of two chains – one chain has 100 scan flops and another chain has 150 scan flops. Patterns can be shifted on both these chains simultaneously. In the second chain, 150 cycles would be required for value to be propagated till the end of the chain. Out of these, 100 cycles can be used for simultaneous shifts in the first chain also. And, the remaining 50 cycles are a waste – from the first chain's perspective. Contrast this with another situation, where the same 250 scan flops are stitched into two chains, each of 125 scan flops. Here, it will take 125 cycles for the values to reach all the scan elements in the design. So, the unbalanced chains scenario required 25 extra cycles, compared to the balanced scan length. Thus, having a balanced chain length prevents any wastage of clock cycle.

6.6 Pseudo Random Flop Outputs

As the values are shifted through the scan chain, the flops along the chains store values as determined by the ATPG patterns. If you look at the design from functional aspects, it is as if these flops are taking random values. Because of the flops taking random values in the Shift mode, there could be several issues – that need to be protected against.

6.6.1 Tristate Controls

Say, a signal is being driven by several tristates. During normal mode of operation, the tristate controls are always controlled in a manner such that only one driver is ON at any given time. But, during *scan-shift*, the tristate enables are getting values randomly (if the enable is being generated from flop outputs). So, it is possible that multiple tristates might get turned ON simultaneously. If this happens, there is a strong likelihood that the device might get burnt during testing. Hence, tristate enables have to be controlled through some primary inputs – in testmode, which will allow ATPG engines to not worry about multiple tristate drivers being turned ON simultaneously – as it generates pseudo-random patterns. One way of achieving this is shown in Fig. 6.11.

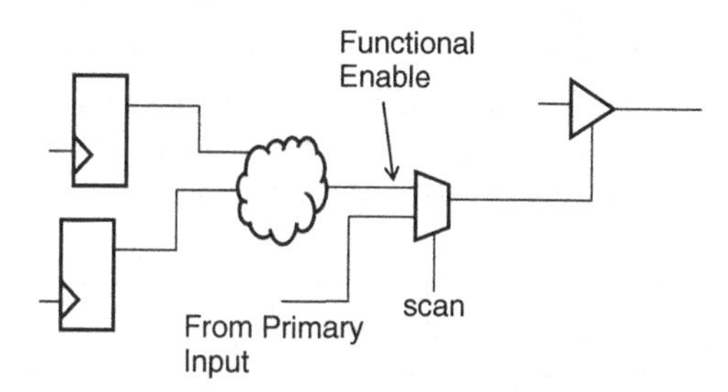

Fig. 6.11 Tristate enable control

6.6.2 Uncontrollable Flops

Flops need to have all their control inputs being directly controllable by ATPG tool. These control inputs include *clocks* and *asynchronous set/reset*. Flops that do not satisfy these criterions are called *uncontrollable flops*. These control signals may not be directly controllable if they are being generated internally within the design. As the signals are being shifted through the chains, these control signals (being generated internally) can randomly assert or deassert. Hence, these flops are called *Uncontrollable*. An assertion of any of the asynchronous control signals will corrupt the value on the flops. Similarly, if a clock (being generated internally) gets missed, it will not shift the value in to the corresponding flop. Thus, the randomness on these internally generated control signals can corrupt the entire chain. Hence, such flops cannot be inserted into the chain. This reduces the testability of the design. Alternately, you have to put bypass mechanisms – so that in scan mode, there is a direct control on these clocks or asynchronous control. This direct control through a primary input allows the specific flop to be included in the chain.

6.6.2.1 Asynchronous Controls

Assume, a flop's asynchronous control such as *set* or *clear* is being generated internally within the design, using combinational logic from flops. As patterns shift across scan chains, these control signals could get asserted, thereby altering the values on the flops – thus corrupting the chain. Hence, such asynchronous controls have to be controlled through primary inputs in testmode. This is achieved using a mux – similar to the concept shown in Fig. 6.11.

6.6.2.2 Clock Gating

The *enable* signal of a clock gate could also be generated internally within the design, based on output of various flops. As the scan patterns are being shifted through the chain, the enable might get turned off in some cycles, thus, disabling the clocks in those cycles. This will cause the scan chain to be corrupted. Hence, the enable signal of the clock gating cell needs to be bypassed in scan shift mode – so that it is ON during *shift*. You have already seen the *Integrated Clock Gating Cell* in the previous chapter. Figure 6.12 shows the same *Integrated Clock Gating Cell* with modifications to account for scan shift also.

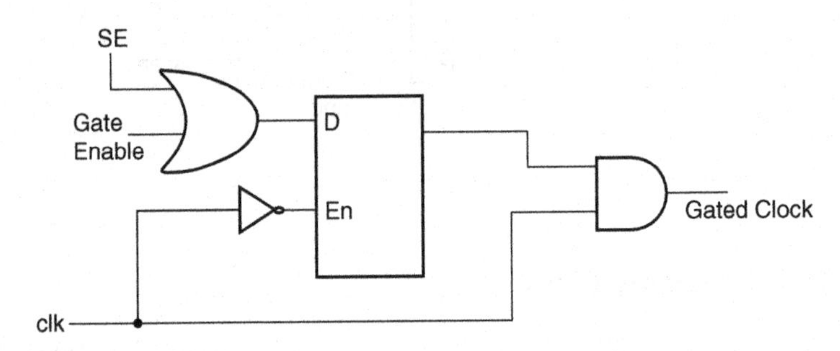

Fig. 6.12 Integrated clock gating cell with scan control

6.7 Shadow Registers

6.7.1 Observability

Figure 6.13 shows the input side of an *Integrated Clock Gating Cell* once again.

As discussed in the previous section, while scanning, the *Gate Enable* is bypassed (in order to make the *Gated clock* to be controllable). This also means that the combinational cloud *C1* gets bypassed during *scan* mode. So, this cloud cannot be tested.

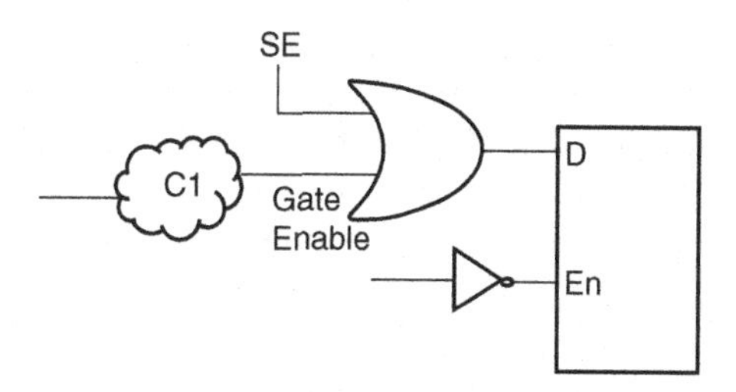

Fig. 6.13 Input side of integrated clock gating cell

In order to test this cloud, a *shadow register* is inserted. *Gate Enable* is fed to the *SI* input of a scan flop. This scan flop is used only in scan mode. During normal (functional) mode of operation, this flop does not have any role. This flop is part of the scan chain. So, in order to test the combinatorial cloud, the value at the output of the cloud is captured into this *shadow register*, and, this captured value can be shifted out. Since the cloud itself is bypassed, so, its value does not cause any gating of the clock – during scan. This is an example of a *Shadow Register* used for improving the *observability*. This shadow register allowed you to observe the values at the output of *C1*, which was otherwise not being observable. You need to put such shadow registers wherever the design has issues with observability. Obviously, all the circuits which are bypassed during scan fall in this category.

6.7.2 Scan Wrap

Consider a memory. In order to observe the values reaching its input, the values would need to be written into the memory, and then read back on the output port side. And these read values would need to be scanned out. Similarly, in order to control the values on the output port side of the memory, the required values would first need to be scanned into the input side of the memory, then written into the memory and then read on the output side. Memory access is very slow, compared to the rest of the system. And so, the entire process will become slow, if you have to involve memory read andwrite also into the testing mechanism. Hence, the memory is surrounded by a set of shadow registers as shown in Fig. 6.14.

F1 represents a set of shadow registers, each of which captures a single bit on the input side of the memory. *F2* represents a set of shadow registers, each of which is multiplexed with the corresponding bit of the memory output port. The sets of

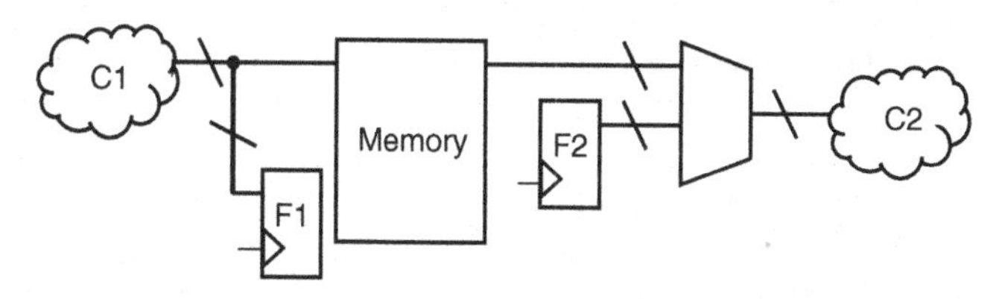

Fig. 6.14 Scanwrap around memory

registers *F1* and *F2* are part of the scan chain. Set *F1* acts as observability points for the combinatorial logic *C1*. Set *F2* acts as controllability points for the combinatorial logic *C2*. This wrapping of the memory by a set of scan flops is called *Scan Wrap*. It should be understood that the *Scan wrap* does not impact the memory's own testing. Rather it improves the *observability* of the cloud on the input side, and, the *controllability* of the cloud on the output side.

6.8 Memory Testing

Memories are more likely to have defects, because of bit-cell architecture. Besides, memories are very densely packed structures. Hence, they are more likely to suffer from shorts, either with power, ground or with adjacent cells. Hence, for memory testing, a set of *Marching Patterns* are used. A set of *0s* need to be written at all locations. Then, all the locations need to be read back to confirm that they are all *0s*. This confirms no *stuck-at 1*. Now, a set of *1s* are written at all locations. Then, all the locations need to be read back to confirm that they are all *1s*. This confirms no *stuck-at 0*. Now, alternate patterns of *0s* and *1s* are written. e.g. assume a memory of 8 bits word. At the first word, *10101010* is written. At the next word, *01010101* needs to be written, and so on. These patterns with alternate *0* and *1* are also called *checkered patterns*. Once the entire memory has been written into, all the locations need to be read back – to confirm that the memory is able to faithfully reproduce whatever was written into it. This confirms against a short with any of the neighboring cells.

In today's modern designs, there are lots of memories. And, trying to test all the locations by reading and writing into it will require a huge number of clock cycles. And, sending all these clock cycles through the tester equipment would mean a huge amount of time. Hence, memory test needs to be put directly on the chip. The chip itself has to generate all these patterns, the associated clocks, the corresponding addresses, read, write and other control signals. And, it also needs to validate the values read back from the memory. Finally, the chip should simply produce a good or bad decision. This is called *Built In Self Test* (*BIST*). Memories are usually tested through *BIST* mechanism, rather than through scan mechanisms on the tester. So, if your design has memories, you also need to put in the corresponding BIST structure.

6.9 Latch Based Designs

Designs with very high performance requirements use latches instead of flops. For such designs, latches are used instead of flops for the scan chains. Consider such a latch based design, where, the latches are stitched into a chain, as shown in Fig. 6.15.

Say, when a positive edge of the clock arrives, *SI* moves into *L1*. However, while the clock is high, this signal can continue traversing through *L2*, *L3* and so on. Depending upon the delays on the scan path, it is not known, how many such elements will be crossed by the signal. A chain with such unpredictability cannot be

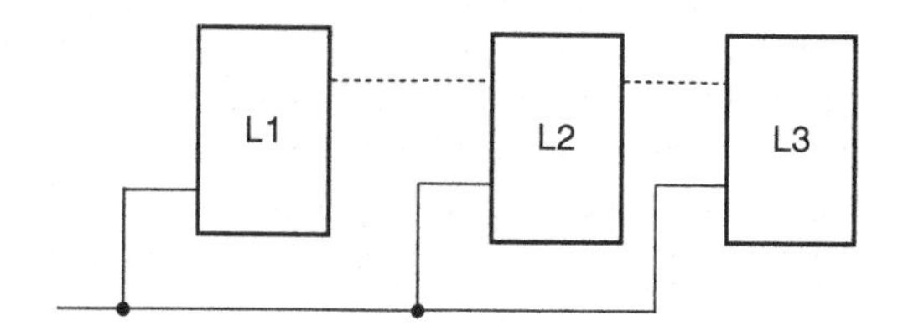

Fig. 6.15 Latches put into a chain

used for scan testing. Thus, for latch based designs, *Level Sensitive Scan Devices* (*LSSD*) are used. Figure 6.16 shows the working of an *LSSD*.

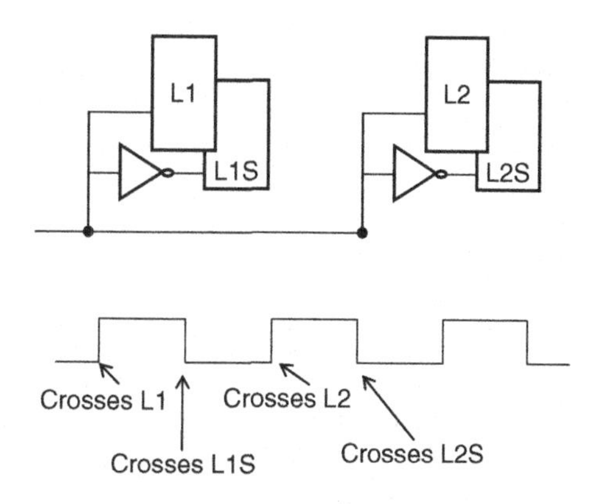

Fig. 6.16 Working of an LSSD

L1 and *L1S* together form one *LSSD* cell. An *LSSD* may be thought of as two back-to-back latches, which are transparent on complimentary phases of clocks. So, when clock has a high, *SI* will move across *L1*, but, will not be able to cross *L1S*, till clock goes low. And, when the clock goes low, it can cross *L1S*, but, it will not be able to cross *L2*. So, in one pulse of clock, it can cross only one *LSSD* device. This brings predictability into the scan mechanism. You now know that the number of scan devices crossed would be exactly equal to the number of clock pulses applied.

For individual latches that are not converted into LSSD, they should be kept transparent in scan mode. A simple method to achieve this is by *ORing* the scan enable signal with the latch's functional enable, and, using the output of this *OR* – for the enable of the latch, as shown in Fig. 6.17.

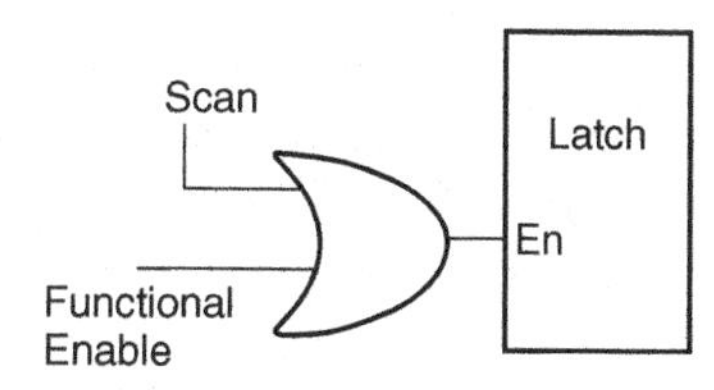

Fig. 6.17 Making latch transparent during scan

It is exactly because of the same reason (ensuring that always only one scan device is crossed in one clock pulse) – that positive and negative edge triggered flops are not interspersed into the same scan chain. All the negative edge triggered flops are put either at the beginning of the chain, or, at the end of the chain. But, they are never interspersed with positive edge triggered flops.

6.10 Combinational Loops

Combinational loop in a design also creates its own challenges for DFT. Consider a loop as shown in Fig. 6.18.

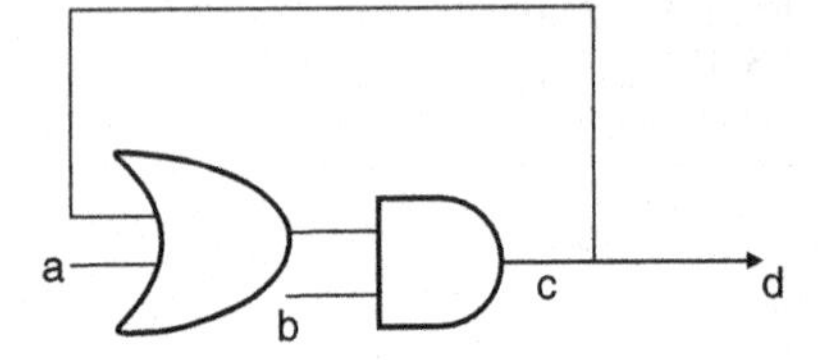

Fig. 6.18 Combinational loop

If we keep *b* at *1* and *a* at *0*, the value of c will keep on circulating. Even a transient *1* at *c* will cause *c* to be stuck-at *1*. Actually, a transient *1* at *c* (or *a*) will keep on circulating if *b* is kept at *1* (irrespective of the values anywhere else in the loop). Thus, controllability of *c* and its fanout requires that *b* cannot be kept at *1*, which means *b* has to be kept at *0*. By putting *b* at *0*, we are losing on the observability of *a* and its fanin cone. Also, with *b* kept at *0*, *c* cannot be controlled to *1*. Thus, ATPG tools will have to put in much more effort to test the logic around this combinational loop.

One technique that you can use is to break the loop in scan mode. This is achieved by inserting a *mux* in one of the segments, as shown in Fig. 6.19.

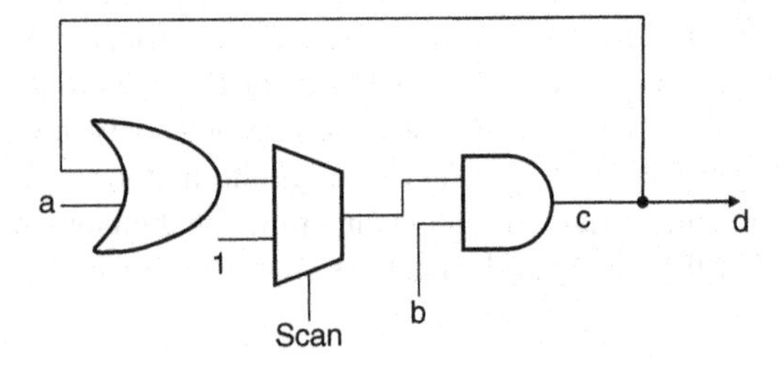

Fig. 6.19 Breaking the combinational loop

During scan mode, the mux transmits a *1*. This allows the AND gate and its fanout cone to be controllable (assuming, there are no other issues around controllability!!). However, now, because of this mux, the fanin cone of *a* has become unobservable. This is resolved by putting a shadow register at the output of the *OR* gate, which can act as a scan flop. A loop might be inherently present in the design – due to the requirements of the functionality of the design. It is also possible that the design did not have a loop in its functional mode of operation, but, a loop got formed, because latches had to be made transparent in the scan mode.

6.11 Power Impact

In Chapter 5, you have seen that higher switching activity means higher power. During scan shift, the flops in the chain are all switching (pseudo) randomly. This results in a lot of switching activity in the logic being fed by these flops. Till about a decade back, power during scan was not given much importance, because, scan is usually carried out at lower frequency. However, despite being at low clock frequency, the overall switching activity is so high, that, sometimes, the power consumed during scan might even exceed the power during normal mode of operation. Hence, power considerations during scan have also become very important.

During scan shift, any activity on the functional path is of no importance. Hence, all activity here is simply a waste of power. Thus, a very simple technique to reduce scan power is to put an *AND* gate at the output of the scan flop, on the functional path. The other input of this *AND* gate is the inverse of *ScanEn*. This ensures that during scan shift mode, this *AND* gate always puts out a *0*. Hence, the functional path does not see any switching activity. This helps reduce the power consumption significantly during scan mode. Fig. 6.20 shows the concept.

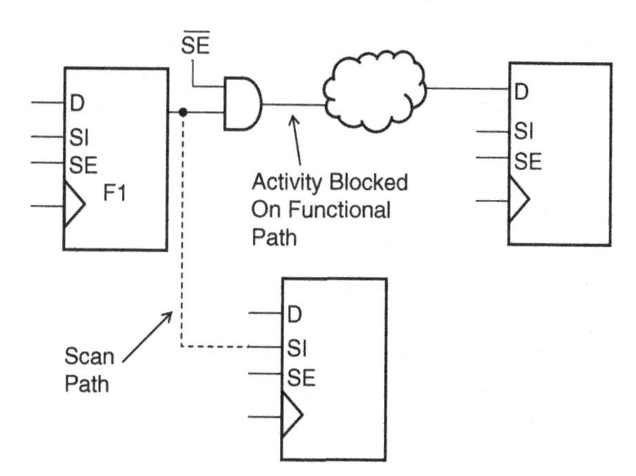

Fig. 6.20 Reduced activity on functional path during scan shift

The cost to this is obviously additional area in terms of these *AND* gates. And also, additional delay in the functional path – due to this additional *AND* gate. In such a case, the *AND* gate might be excluded on the top few critical paths. This ensures that though the power is reduced significantly, but, the operational frequency of the device remains unaffected.

6.12 Transitions Fault Model

Besides the *stuck-at* fault model discussed in this chapter, another model is the *transition fault* model. *Stuck-at* fault testing is carried out at lower frequency. So, these tests can tell you, whether the *gates* are stuck or not. However, they do not test, if a *gate* takes too long to go to its desired value. A *gate* (or a path) taking too long to reach its desired value indicates that the device will not operate at its desired frequency.

In *transition fault* model, the testing is carried out – at the frequency at which the device is expected to perform. The basic concept of *shift* and *capture* is still the same. The only thing additional here is to ensure that the paths switch in the desired time. Here, if a *capture* is not correct, it indicates that something on that path takes longer to switch than the expected time, which means, this device will not be able to operate at the desired frequency, though, it might still be able to operate at a lower frequency.

The *stuck-at* fault model provides a good or bad kind of decision, in the sense that if some device fails on the tester, it indicates the device is no good in terms of its functionality also. However, if a device passes the *stuck-at* tests, but, still fails on the *transition* fault, it might not necessarily be a case of rejecting the device. The device might still operate but at a lower frequency. Depending upon the application, it is possible that such devices with lower frequency might still get used in a lower-cost application!!! These tests are also called *at-speed* tests, because, these are carried out at the same frequency as the frequency of actual operation of the device.

For lower geometries, the frequencies are much higher. That means each gate in the path has much smaller delay at its disposal. Thus, if a gate is slow to switch, it might fail the timing. This explains why *at-speed* tests are becoming very prevalent for advanced technology nodes. At-speed tests post much more complex challenge to the ATPG tools. However, in terms of design also, they have additional burden on the design of the clock network. The PLL has to generate exactly two pulses (one to launch and another to capture) of the functional clock, after the pattern has been shifted into the respective scan elements. Also, while flops belonging to different clock-domains can be put into the same scan chain, as long as the test clock is the same for regular scan testing; the same is not true for at-speed testing. During at-speed testing, each of the flops will be driven by the system clock. And so, if two elements in a chain are operating on asynchronous system clocks, they will not be able to ensure a reliable capture. Multicycle paths (explained later in Chapter 7) also pose a major challenge in at-speed testing.

6.13 Conclusion

In order to make it easier for the device to be tested against any manufacturing defects, you have to ensure that all parts of your design can be controlled as well as observed. One of the most major requirements is that all control signals (clocks that trigger any flop, asynchronous signals which can set or reset any flop, clock-gating signals, tristate controls etc.) should be controllable from outside – directly by the tester. Effectively, DFT techniques make your design easier to test. In other words, you have increased the testability of the manufactured device, by taking certain cares during the design stage itself. Hence, some engineers refer to DFT as *Design For Testability*. Having latches, combinatorial loops etc. in the design cause additional challenges during testing; unless, special care has already been taken during design stage itself. Adding all these testing considerations to the design takes its toll on area, timing and power – however, some cost on these factors is still justified in terms of the tester time that gets saved. Sometimes, on specific critical paths, you might want to use your judgment as to whether to gain on the timing – by compromising with the DFT requirements. On non-critical paths, a slight additional cost in timing is anyways of no concern.

Chapter 7
Timing Exceptions

As seen in Chapter 3, most of the paths in a design need to meet certain timing requirements – specifically, that they should be captured at the destination at the immediately following active edge of the clock, and, should not interfere with the previous active edge of the clock on the destination. However, there are certain paths which need not follow the above requirements. In subsequent sections, you will see various situations and examples of why the timing on certain paths need not meet the single cycle requirement. These paths are called *Timing exceptions.* There are three kinds of exceptions:

- False Paths: These are paths that need not be timed.
- Disable Timing: These are specific segments of a path that are disabled. Thus, any path through that segment will not be timed.
- Multicycle Paths: These paths allow more than one cycle for the signal to reach the destination.

For such paths, it is better to provide the required exceptions. In the absence of exceptions, the following disadvantages might occur:

- If the timing on these paths is relatively hard to meet, the implementation tools will unnecessarily spend too much time – in trying to meet the timing.
- As these tools try to meet the timings for these paths also (along with all other paths), they might deteriorate the timings on other paths – which were really of interest.
- In order to meet the timings on these paths, they might put higher drive cells or buffers which unnecessarily increase both area and power for the device.

7.1 False Paths

False paths refer to those paths, which seem to be structurally existing or connected, but, need not be timed because functionally they don't interact. There might be several reasons, why such a path need not be timed. Such paths are declared as false path. The SDC command for declaring such paths as false is:

S. Churiwala, S. Garg, *Principles of VLSI RTL Design*,
DOI 10.1007/978-1-4419-9296-3_7, © Springer Science+Business Media, LLC 2011

set_false_path -from <a set of objects> -through <a set of objects> -to <a set of objects>

Either or several of *–from*, *-through* or *–to* could be absent. There can be multiple occurrences of *–through* in the same command. For either of *-from*, *-through* or *–to*, the set of objects could contain just a single element or multiple elements. A clock name in *-from* means all data launch triggered by that specific clock. Similarly, a clock in the *–to* means all data capture triggered by that specific clock. There are additional qualifiers that can be put, e.g. –setup, -hold etc. These qualifiers say that the false paths apply only to the specific checks or specific edges, rather than all transitions along the specified paths.

7.1.1 False Paths Due to Specific Protocol

Consider the circuit in Fig. 7.1.

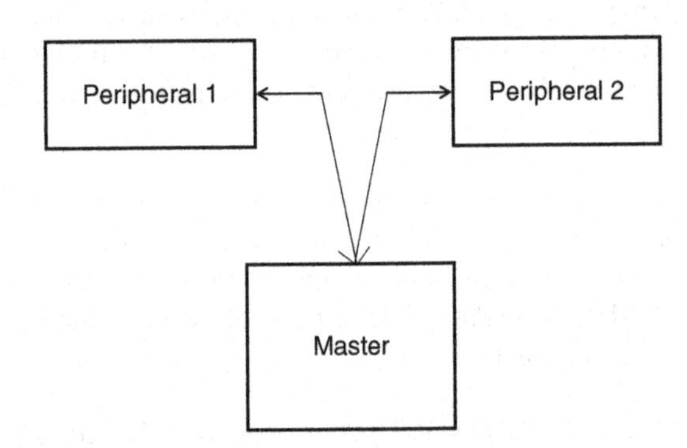

Fig. 7.1 Master communicating with two peripherals

This is an example of a Master which can exchange data (both ways) with two peripherals. Any of the peripherals can send data to or receive data from the master. However, the two peripherals cannot exchange data directly with each other. However, a timing analysis tool will see a structural path between the two peripherals (through the common pin of the master). So, the timing analysis tool will want to check the timing on these paths (peripheral 1 → master's pin → peripheral 2 and peripheral 2 → master's pin → peripheral 1). Such paths need to be declared as False Paths as:

set_false_path -from {Peripheral1, Peripheral2} –to {Peripheral1, Peripheral2}

7.1.2 False Paths Due to Paths Being Unsensitizable

Consider the circuit in Fig. 7.2.

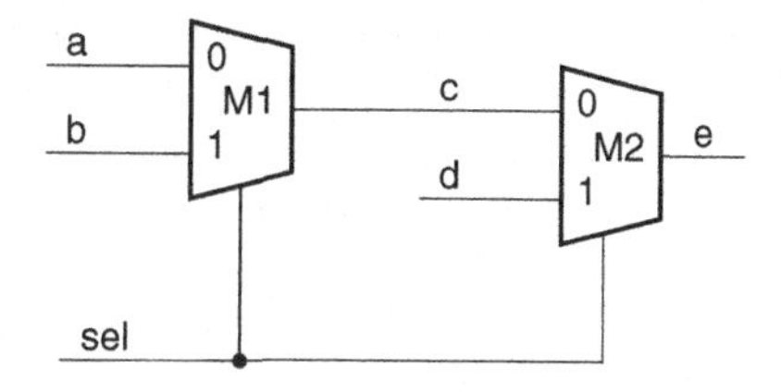

Fig. 7.2 Path "*b*" to "*e*" cannot be sensitized

The path from *b* to *e* cannot be sensitized. Any transition on *b* will not reach *e*. Still, the timing analysis tool will try to time this path also. Hence, the path from *b* to *e* should be declared as a False Path as:

set_false_path –from b –to e

An obvious question that might cross your mind could be: If the path is anyways not sensitizable, why did such a path occur in the circuit? The answers could be several. The most probable being: Synthesis tool ended up inferring this kind of circuit, as part of its optimizations. Maybe, it needed signal *c* – for some other operation. Similarly, another situation where a path cannot be sensitized can occur when some signals re-converge. It is possible that during re-convergence, opposite transitions cancel each other. Thus, the transitions die down at the point of re-convergence. An extreme example of such a situation is shown in Fig. 7.3.

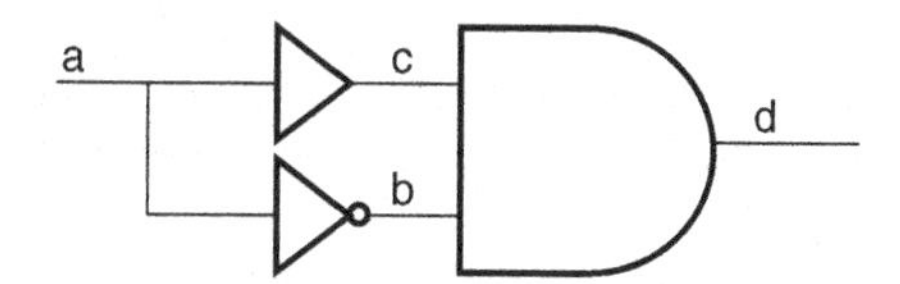

Fig. 7.3 Transitions on "*a*" do not reach "*d*"

Here, any transition on *a* will die down at *d*, because there are two paths from *a* to *d* and they have mutually-opposite unateness. So, theoretically, the path from *a* to *d* can be declared as a false path. The example shown in Fig. 7.3 is mostly for the sake of conceptual understanding only. Usually, reconvergence would not happen in such a simplistic manner. There could be several reasons for such reconvergence being created – with the most probable being: Synthesis tool ended up inferring this kind of circuit, as part of its optimizations.

However, one should be very careful while declaring such paths as False. By declaring these paths as False, a user is effectively allowing this path to have any delay value. Though, the signal at *d* is expected to be held at *0*, there is always a possibility of a glitch at *d* – due to differential path delays from *a* to *c* and from *a* to *b*. If the path is not timed, it is possible that the glitch occurs just at the time, when *d* is captured. In such a scenario, the glitch value will be captured – resulting in erroneous behavior of the device.

One mechanism that you can use is to compute the actual delays of the paths, and then, based on the differential delays at the point of re-convergence, check, if there is a possibility of a glitch. If you are relying on the computation of differential delays, you should remember that anytime there is even a slight modification to the circuit or its layout, a path which was earlier a valid false-path could now generate a glitch, due to a change in the differential delay. It might be better to actually time such paths, rather than take the risk of capturing a glitch.

7.1.3 False Paths Due to CDC

In Chapter 4, you have seen that for an asynchronous clock domain crossing, there is always some combination of clock edges, which will result in a timing violation. Hence, any attempt to time the paths involved in such crossings will result in implementation tools spending too much effort and still not being able to meet the timing. Hence, such crossings need to be declared as falsepaths. One of the most popular ways to declare such paths as false is through:

set_false_path –from <launching clock> -to <capturing clock>

The above command allows for any amount of delay at the point of crossing. Theoretically, this could mean a very high delay. In order to prevent the possibility of a very high delay, many designers prefer to use *multicycle* path, rather than a false path declaration for CDC. Or, some designers put a *set_max_delay* – so that the path is constrained to have some upper limit for delay. SDC version 1.7 has introduced the command *set_clock_groups*. It is better to use this command, rather than using *set_false_path* for CDC. The command for *set_clock_groups* to be used is:

set_clock_groups –asynchronous –group <launching clock> -group <capturing clock>

In terms of timing the paths, both the commands *set_false_path* and *set_clock_groups* have the same impact, i.e. do not time the path (involved in the crossing). However, *set_clock_groups* captures the intent correctly. Also, cross-talk analysis treats both these commands differently. The impact of putting a *multicycle* path is that there is an upper bound created – for the delay for the signal taking part in the crossing. In general, this might be considered as a good safe-guarding practice, but, because anyways there is not much logic on the signal taking part in the CDC, the need for this upper-bound is not really severe.

The commands mentioned here are only from the timing analysis perspective. The need to synchronize and other reliability assurance measures as mentioned in Chapter 4 of the book continue to exist – which you have to ensure. Because the paths are not being timed (or, the timing has been relaxed through multi-cycle declaration), from the timing analysis perspective you will not get any indication even if the synchronization etc. is not done properly.

7.1.4 False Paths Due to Multi Mode

Consider the circuit as shown in Fig. 7.4.

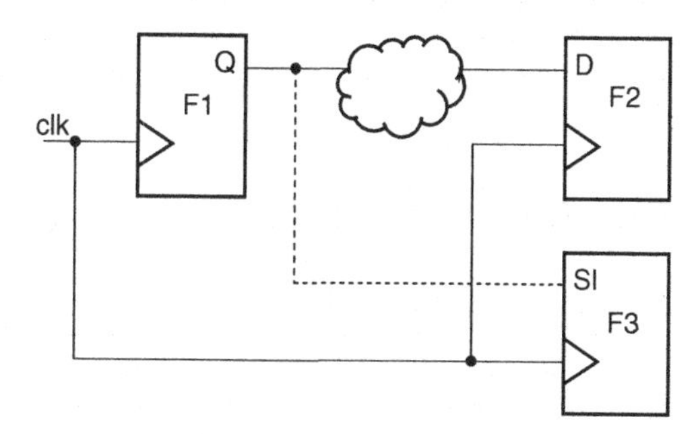

Fig. 7.4 Scan path and functional paths

All the three flops (*F1*, *F2* and *F3*) are being driven by the same clock pin. The path from F1 to F2 is a functional path; and the path from F1 to F3 is a scan only path, which gets enabled in scan shift mode only. During normal mode of operation, the clock operates at a period of 10 (say) – and ScanEnable is kept at 0. So, the constraints are specified as:

create_clock –name FuncClk –period 10 [get_ports clk1]
set_case_analysis 0 [get_ports ScanEn]

And, during scan mode of operation, the circuit is clocked slowly – say with a period of 40. Irrespective of the mode of operation, the clock is still applied at the same point (*clk1*). So, for scan mode:

create_clock –name TestClk –period 40 [get_ports clk1]
set_case_analysis 1 [get_ports ScanEn]

Sometimes, you might want to write the constraints for multiple modes of operation together. It has its own advantages. So, if you combine the constraints for both the modes, they become:

create_clock –name FuncClk –period 10 [get_ports clk1]
create_clock –name TestClk –period 40 [get_ports clk1]

Notice that the *set_case_analysis* is no longer existent. Because of the *FuncClk* declaration, the path from *F1/Q* to *F3/SI* will also get timed at 10, while, actually, this path is desired to be timed at 40 – as it is a scan path. Hence, this path (and, other similar paths) needs to be declared as false as:

*set_false_path –from FuncClk –to *_reg/SI*

The above command says that all paths starting from *FuncClk*, but, reaching the *SI* (ScanIn) pin of all registers need not be timed. This still causes the scan paths to be timed with respect to TestClk, which is at 40. Theoretically, there needs to be a converse also – to exclude the paths from *TestClk* to **_reg/D*. Since those paths are anyways expected to meet the time of 10 (due to *FuncClk*), they will automatically meet the timing of *TestClk*. Thus, many designers don't put the converse.

7.1.5 False Paths Due to Pin Muxing

The physical size of a chip depends on:

- The area within which all the silicon and its interconnects would fit
- The periphery which can accommodate all the pins

As device size is decreasing, the first factor is no longer dominant. Rather the final physical size is getting dictated by the periphery required to accommodate all the pads. This situation is called *Pad Limited*, because, the device size can no longer be reduced – due to the limitations in being able to place the required number of pads. A standard technique in such cases is to do pin-multiplexing, also called pin muxing. On the input side, it means that the same input pin will serve two different purposes during different modes of operation. Similarly, the same output pin will serve two different purposes during different modes of operations. Some of the commonly used situations where such pin muxing can be used include:

- During functional mode a pin is used for some functional data, and the same pin is also used for scan data – during scan mode.
- If a memory takes address and data in two different cycles, then, the same pins can be used to send address in one cycle and data in another cycle.

Consider the circuit shown in Fig. 7.5.

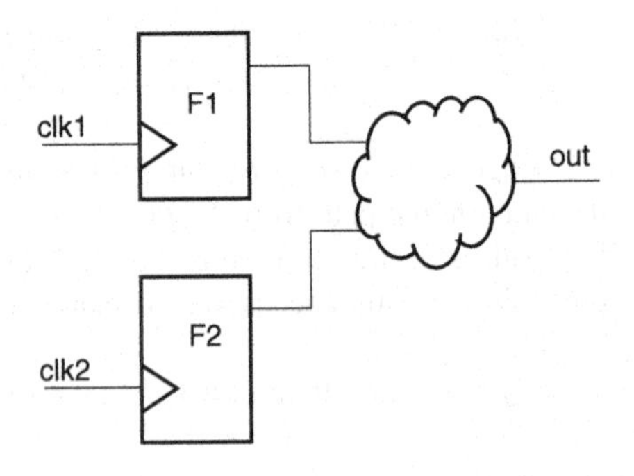

Fig. 7.5 Output pin muxing

Depending upon the mode of operation, *out* will get either the data from *F1* or from *F2*. There is no mode, where both *F1* and *F2* can send the data onto *out* simultaneously. Both *F1* and *F2* are clocked by two different clocks. When the data coming out of *out* is being launched by *F1*, it will also be captured outside by the same clock: *clk1*. Similarly, the data launched by *F2* will be captured outside by the same clock: *clk2*. Hence, there needs to be two output delays on *out* as:

set_output_delay <value> -clock clk1 [get_ports out]
set_output_delay <value> -clock clk2 [get_ports out] –add_delay

During timing analysis, the following 4 paths get timed:

1. Data being launched by *F1* and sampled by *clk1* (due to the first set_output_delay)
2. Data being launched by *F1* and sampled by *clk2* (due to the second set_output_delay)
3. Data being launched by *F2* and sampled by *clk1* (due to the first set_output_delay)
4. Data being launched by *F2* and sampled by *clk2* (due to the second set_output_delay)

However, the paths on interest are only the (1) and (4). The paths (2) and (3) are not of interest – because, they are timing a situation that will never really happen.

One simple solution to this is to write false paths between *clk1* and *clk2* (both ways), such as:

set_false_path –from clk1 –to clk2
set_false_path –from clk2 –to clk1

The first false path declaration prevents the timing of (2), because, *F1* is triggered by *clk1*. And, the second false path declaration prevents the timing of (3). However, this creates a risk that any other interaction inside the design amongst *clk1* and *clk2* might inadvertently get declared as false. In order to protect against this risk, you should declare virtual clocks (say: *vclk1* and *vclk2*) corresponding to the clocks *clk1* and *clk2*. Now, the output delay should be specified with respect to these virtual clocks, rather than the real clocks. And, the false path should be specified with the *–from* being the real clock and the *–to* being the virtual clock, as:

set_output_delay <value> -clock vclk1 [get_ports out]
set_output_delay <value> -clock vclk2 [get_ports out] –add_delay
set_false_path –from clk1 –to vclk2
set_false_path –from clk2 –to vclk1

In case of input side pin muxing, the same concept needs to be used. The only difference is that the *–from* has to be the virtual clock, and, the *–to* should be the real clock. It is worth mentioning that many times, for pin muxing, RTL designers use two different modes for the analysis. In such a case, the set_case_analysis setting (for the mode selection) will cause the multiplexor to select just one path. The other path would automatically be disabled. So, if you are following the style of multiple modes, you would not need these false path declarations.

7.1.6 False Paths Due to Exclusive Clocks

Consider the circuit shown in Fig. 7.6.

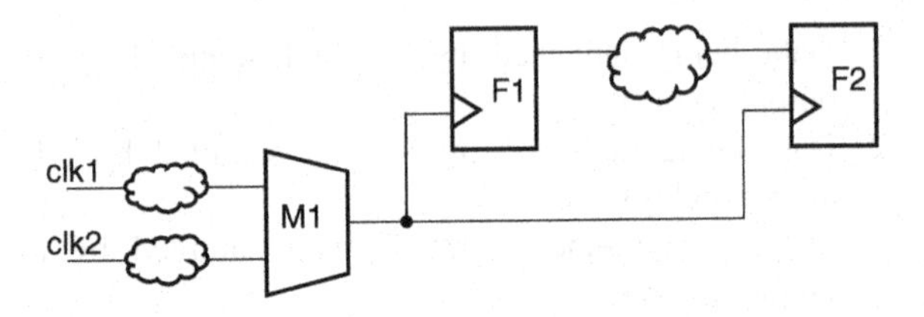

Fig. 7.6 Clocks *clk1* and *clk2* are exclusive

Both flops *F1* and *F2* can get either *clk1* or *clk2* clock. However, it is not possible to have a situation, where, *F1* gets one of these clocks, and, *F2* gets another of the clocks. This is because the mux *M1* selects only one among these clocks. Thus, there is no need to time for the following situations:

- *F1* launching data on *clk1* and *F2* capturing the data on *clk2*
- *F1* launching data on *clk2* and *F2* capturing the data on *clk1*

This can be achieved simply by the false-path specifications (both ways) as:

set_false_path –from clk1 –to clk2
set_false_path –from clk2 –to clk1

Some STA tools allow variables which can be set to consider only one clock among the several ones reaching the registers. However, it is better not to depend on such variables because:

- The tools do not define which of the multiple clocks would be chosen for the analysis involving that register.
- All STA tools might not necessarily support this variable; and hence, depending upon the setting of this variable would make your constraints and design to be less portable across different methodologies or tools.

Also, with the introduction of set_clock_groups in SDC1.7, this kind of situation is better expressed as:

set_clock_groups –physically_exclusive –group clk1 –group clk2

Irrespective of which of the above two commands is chosen, there is still a risk. Consider the circuit shown in Fig. 7.7 – which is a slightly modified version of the Fig. 7.6.

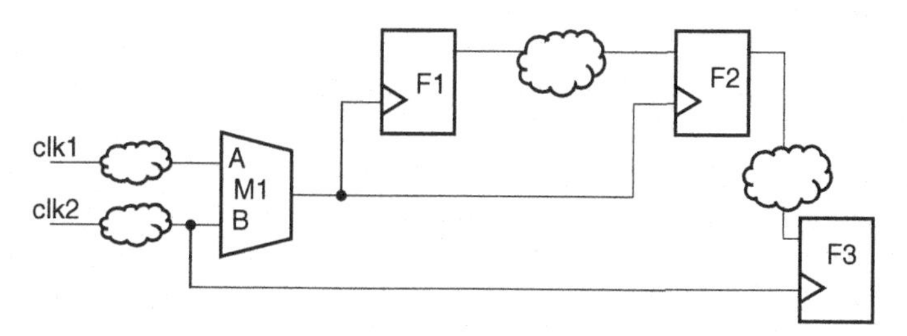

Fig. 7.7 Clocks *clk1* and *clk2* are no longer exclusive

Now, it is possible that while *F2* launches data on *clk1*, *F3* captures the data on *clk2*. This path will now not be timed, even though it is a valid path, thus, causing a risk of timing failure of the device. Hence, while declaring such *set_false_path* or *set_clock_groups*, it should be checked that the clocks are not directly reaching any of the sequential elements, before reaching the multiplexing element. Further, these checks have to be made after each change to the design, just in case an enhancement or modification has introduced this additional flop – represented in Fig. 7.7 by *F3*. A simple method to do away with this risk is to have generated clocks declared at the input pins of the MUX. These generated clocks (say: *gclk1* and *gclk2*) should have its source and master as *clk1* and *clk2* respectively. Now the false path or physically exclusive clock groups should be specified with respect to these generated clocks, rather than the actual master clocks.

create_generated_clock –name gclk1 –combinational –source [get_ports clk1] –master_clock [get_clocks clk1] M1/A
create_generated_clock –name gclk2 –combinational –source [get_ports clk2] –master_clock [get_clocks clk2] M1/B
set_clock_groups –physically_exclusive –group gclk1 –group gclk2

7.1.7 False Paths Due to Asynchronous Control Signals

Asynchronous control signals – which are typically kept asserted for several cycles and control almost whole of the sub-system or system might need to be declared as false-paths. Since they are asynchronous – they will impact as soon as they reach the element; also since, they are typically held active for multiple cycles, it is expected that the asynchronous control will reach all the desired elements. Hence, there is no real need to time them.

set_false_path –from rst_n

However, though, the assertion of the signal is asynchronous, for de-assertion it is important to have all the elements come out of reset at the same time. Hence, the need to not time is only for the assertion portion. Thus, the above command has to be modified as:

set_false_path –from rst_n –fall (assuming, the reset is active Low)

Even for the assertion, putting a *set_false_path* gives a complete freedom to the implementation tools to take as much delay as they want. To avoid this, many designers prefer to put an upper-bound either through *set_multicycle_path* or through a *set_max_delay*.

7.1.8 False Paths Due to Quasi Static Signals

Certain signals are not expected to change in the middle of an operation. They are expected to be set once – and then, continue to retain their value – in that specific mode of operation. Since these signals do not change their values during the middle of an operation, these need not be timed. Paths along such signals can also be declared as false. Typical examples could be configuration registers, test-mode etc.

7.1.9 set_false_path -vs- set_clock_groups

You have seen that for certain situation involving paths between different clocks, you can use either *set_false_path* or *set_clock_groups*. In terms of impact on whether or not the path is included (or excluded) for timing-analysis, the results are the same, irrespective of which of these commands you use. However, it is more appropriate to use *set_clock_groups*. Use of *set_clock_groups* has the following advantage over the use of *set_false_paths*:

- It communicates the intent more clearly and explicitly.
- It is more specific in terms of the reason why the paths between a clock pair are not to be timed. Whether, the clocks are asynchronous, or, logically exclusive or physically exclusive; rather than *set_false_path* which simply tells that the paths should not be timed.
- The treatment of noise or cross-talk analysis is different. *set_clock_groups* provides the right treatment for cross-talk analysis.

However, because *set_clock_groups* has been a recent addition to SDC, many of the old constraints still use *set_false_path*.

7.2 Disable Timing

Sometimes, certain specific arcs within a given path have to be broken. The most common reason for this is the presence of a combinational loop. Consider the circuit shown in Fig. 7.8.

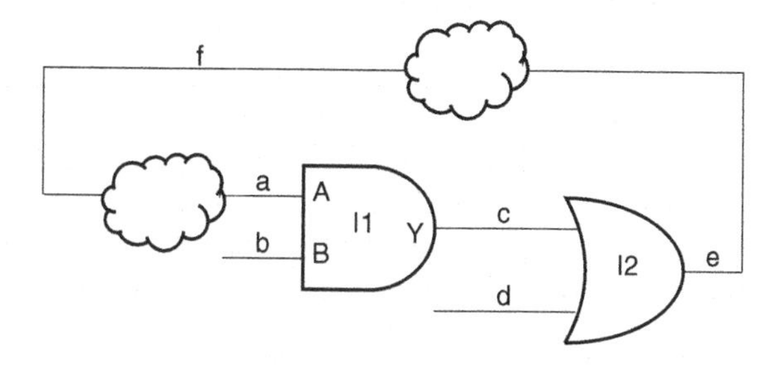

Fig. 7.8 Combinational loop

A combinational loop is formed through the signals *a, c, e, f,* and back at *a*. The delay through an element gets impacted by its input transition time. And, the transition time at the output of an element is also a function of the transition time at its inputs. Consider a transition at *b*. This starts the path delay and transition computations at *c* and then at *e* and then at *f* and then at *a* and then again at *c*. The transition time at *c* impacts the transition time at *a*, via the transitions times at *e* and *f*. The new transition time at *a* once again impacts the transition time at *c*, and, so on. Thus, for a timing analysis tool, this computations would keep on iterating. Usually, all timing analysis tools are capable of dealing with this situation by breaking one segment of the loop. Many designers prefer to break the loop explicitly – by disabling one of the arcs. Looking at instance *I1*, you can break the loop by the command:

set_disable_timing I1 –from A –to Y

The advantage of explicitly breaking the loop is that the designer knows which segment is being broken. This is especially important, because, when a segment is broken, it breaks the loop alright, but, it also prevents timing through all the paths which include that segment. Consider the circuit shown in Fig. 7.9.

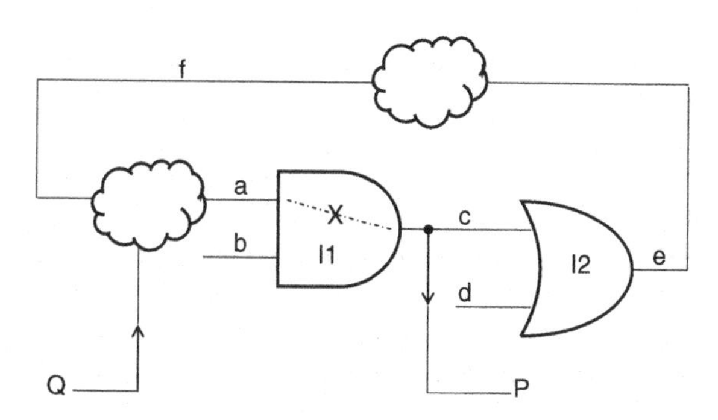

Fig. 7.9 Segments of a loop impacting other paths

As soon as the above *set_disable_timing* is specified, it also results in the path *Q, a, c, P* not getting timed. Given the circuit of Fig. 7.9, a better place to break the loop could be on *I2*, or, on any other segment – that is not a part of any other path.

7.3 Multi Cycle Paths

Sometimes, it is not necessary for a launched data to be captured at the destination in the immediately following edge. In such cases, the data maybe allowed to take more than one clock cycle to reach the capturing device. Such situations are called *Multicycle*. So, effectively, a *Multicycle Path* specification tells STA that the time for capturing the data can be extended beyond the normal single cycle. There are various reasons, why a design might have the need for a *Multicycle Path*.

7.3.1 *Slow to Fast Clock Transfer of Data*

Consider a situation, where data is being generated by a slow clock, and is being captured by a fast clock, which is some multiple (in terms of frequency) of the generating clock.

7.3.1.1 Need *for Multicycle* -setup

Consider the circuit shown in Fig. 7.10a.

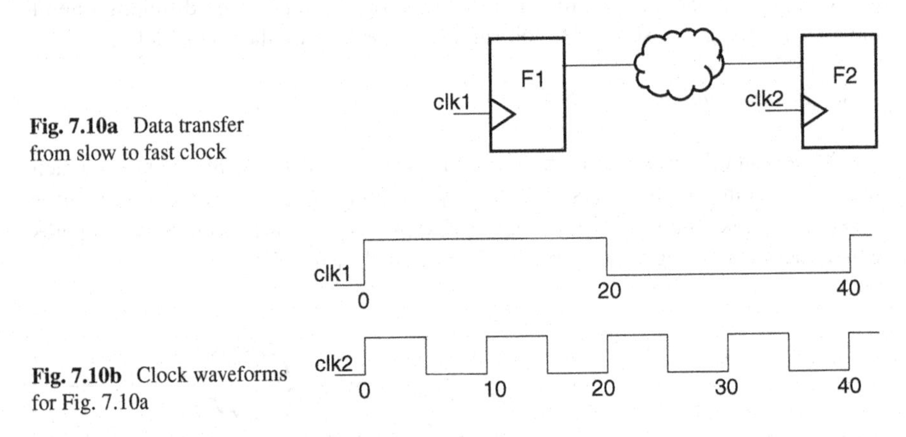

Fig. 7.10a Data transfer from slow to fast clock

Fig. 7.10b Clock waveforms for Fig. 7.10a

Assume the flop *F1* is clocked by a clock (*clk1*) with period 40, and, *F2* is clocked by another clock (*clk2*) with period 10. The corresponding waveforms are shown in in Fig 7.10b. Data is launched by *F1* at time 0. STA will try to check if the data will reach *F2* before the next clock edge on *F2*, i.e. at 10. But, for next several clocks on *F2*, there is no new data – which can overwrite the current data. So, *F2* might as well capture this new data at 40, or, anytime before it.

So, you can declare this path to be a *set_multicycle_path* of 4. This declaration is effectively telling the implementation tools that they can take enough time to have this data reach *F2* even up to 4 cycles (of destination clock), rather than having to rush the data. The corresponding command is:

set_multicycle_path –setup 4 –from clk1 –to clk2

This command is also telling the STA tool to allow up to 4 cycles (of destination clock) for the data to reach the destination flop. Figure 7.11 explains the impact of *set_multicycle_path* specification for *–setup*.

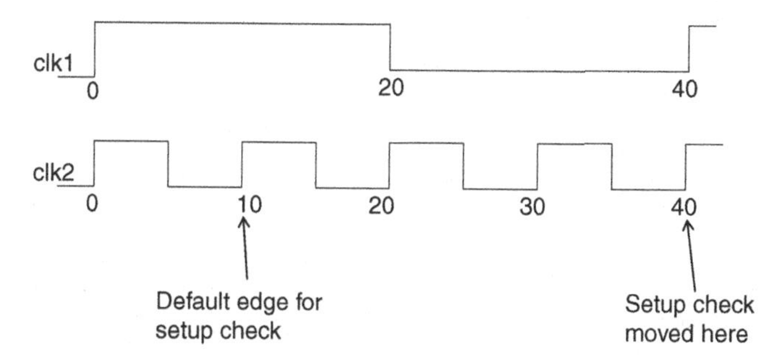

Fig. 7.11 Movement of setup edge

The numbers mentioned in the *set_multicycle_path* command are in terms of destination clock (unless, special qualifiers are used to indicate otherwise). Thus, for any slow clock to fast clock (synchronous) crossing, there might be a *set_multicycle_path –setup N*, where, N is the ratio of the clock periods.

7.3.1.2 Impact on Hold Analysis

The *hold* checks are done with respect to the destination edge immediately preceding the edge used for *setup*. Thus, moving the *setup* edge causes the *hold* edge also to be moved automatically. So, with *set_multicycle_path* being specified as 4, the *setup* check is being made at 40. That means, the *hold* check is being made at 30, which is the immediately preceding edge. In Section 3.4.3, you have seen that *hold* check means data should remain stable after the active edge of the clock. Thus, a *hold* check at 30 implies that the launched data should not reach the destination flop anytime before 30. So, effectively, the overall impact has been: data has to reach *only* between 30 and 40. But, this is not what was intended. The intention was to provide a flexibility that data can reach *anytime till* the 4th clock edge. What you ended up conveying is that data should reach ONLY in the 4th cycle. If the actual delay for the path was anyways coming out to be less than 30, then, this *hold* requirement (at 30) will cause additional delay elements to be inserted, which will unnecessarily take up silicon area as well as power.

So, you need to move the *hold* check back to 0 – from 30. That means, the *hold* check has to be moved back (towards 0 – away from the *setup* edge) by 3 cycles. This is achieved by completing the specification as:

set_multicycle_path –setup 4 –from clk1 –to clk2
set_multicycle_path –hold 3 –from clk1 –to clk2

This specification is effectively telling that the *hold* check edge is moved towards origin by 3 cycles. So, the *hold* edge has moved back to its original location. Figure 7.12 explains the movement of *hold* edges.

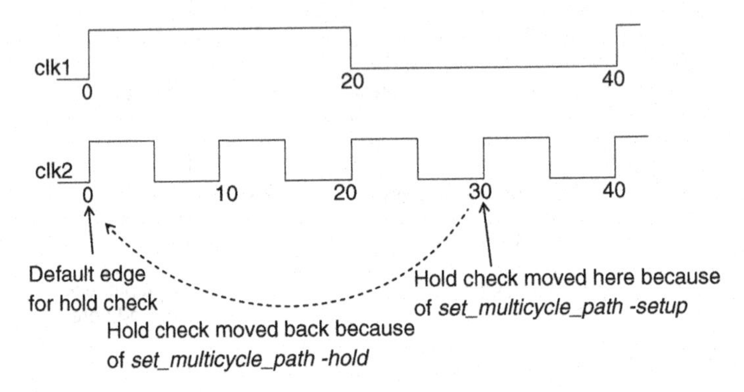

Fig. 7.12 Movement of hold edge

Now, you have achieved what you had wanted to convey, viz: the signal can reach *F2* anytime until 40.

7.3.1.3 Protection Against Glitch Capture

Assume that the actual path delays are such that at time 10 (or 20 or 30), there is some glitch at *F2*. At these times, since *F2* is getting clocked, so, it might capture glitches. This could result in a functional failure of the device. So, the design has to ensure that *F2* should not really capture anything before 40, irrespective of when the data reaches it. This is a very important precaution that you have to take in your design. One simple method of achieving this could be as shown in Fig. 7.13.

The comparator output becomes 1 only once in every fourth cycle (of *clk2*). So, *F2* samples the new data only in the 4th cycle. For the previous 3 cycles, it simply recirculates its own value. Hence, even if there is any glitch or transient value on the data line (coming from *F1*), that glitch would not be captured. Another alternative approach to avoid a glitch is ofcourse to not provide the *–hold* specification. This will prevent the signal to be updated anytime before the 4th cycle. Since signal can not change in earlier cycles, hence, there can be no glitch. But, as explained in Section 7.3.1.2, this could cause unnecessary buffers to be inserted to forcefully increase the delay to be more than 3 cycles.

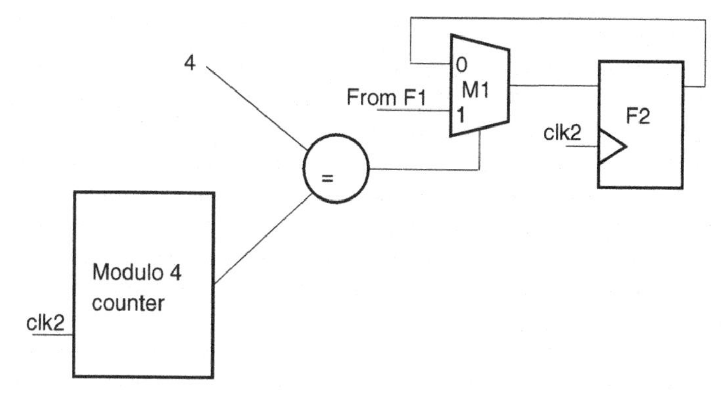

Fig. 7.13 Avoiding glitch capture

7.3.2 Fast to Slow Clock Transfer of Data

Consider a situation, where data is being generated by a fast clock, and is being captured by a slow clock, which is some division (in terms of frequency) of the fast clock.

7.3.2.1 Need *for Multicycle* –setup

Consider the circuit shown in Fig. 7.14a.

Assume the flop *F1* is clocked by a clock (*clk1*) with period 10, and, *F2* is clocked by another clock (*clk2*) with period 40. The corresponding waveforms are shown in Fig. 7.14b. Here, the data is being generated at a rate which is faster than the rate at which it is being captured. Hence, there will be data loss. In order to prevent the data loss, there has to be mechanism in place controlling *F1*, so that it does not really generate the data at each edge of its clock. Rather, it should generate data only once

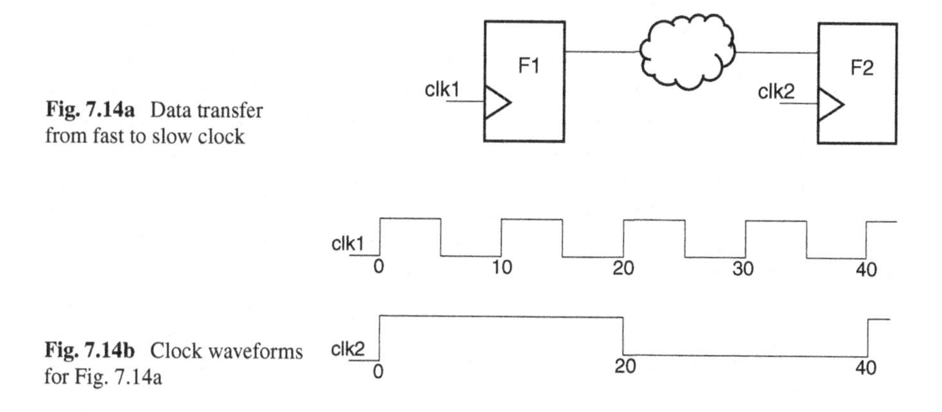

Fig. 7.14a Data transfer from fast to slow clock

Fig. 7.14b Clock waveforms for Fig. 7.14a

in one cycle of the destination clock (i.e. *clk2*). So, *F1* can send a new data only once among the 4 edges (0, 10, 20 and 30).

If *F1* generates the data at 30, the data gets only 10 time units to reach *F2* at 40. So, *F1* might as well generate the data at 0 – rather than at 10 or 20 or 30. This will allow data to get longer duration to reach *F2*. During *setup* check, STA will do the most pessimistic analysis. And, hence, it will assume that the data was launched at 30, and, has to be sampled at 40. But, actually, the data has been launched at 0. So, the launch edge has to be moved back to 0. This effectively means, moving the launch edge back by 3 cycles, where, 3 cycles is in terms of the launch clock. This is done through the command

set_multicycle_path –setup 4 –from clk1 –to clk2 –start

The *–start* in the command says that the movement has to be in terms of the clock on the start point of the path (i.e. *clk1*). Notice that the number specified is 1 more than the number of cycles by which you have to move back the start edge. Fig. 7.15 explains the impact of this *set_multicycle_path* specification.

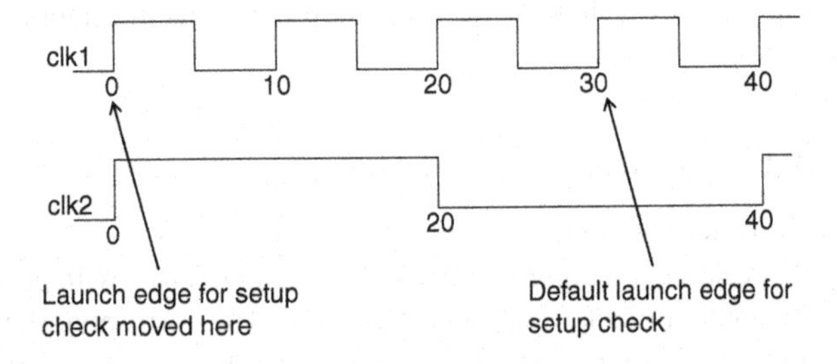

Fig. 7.15 Movement of launch edge for setup check

Thus, for any fast clock to slow clock (synchronous) crossing, there might be a need to specify *set_multicycle_path –setup N -start*, where, N is the ratio of the clock periods.

7.3.2.2 Impact on Hold Analysis

With the *setup* check using the launch edge at 0, the hold check will assume the launch edge at 10, and, the capture edge would still be 40. That means, the *hold* requirement of 30 (40–10) gets created. A *hold* requirement of 30 implies that a new data should reach the destination flop only after a delay of 30. So, effectively, the overall impact has been: data has to reach *only* between 30 and 40. But, this is not what was intended. The intention was to provide a flexibility that data can reach *anytime till* 40. What you ended up conveying is that data should reach ONLY in the time range 30–40. If the actual delay for the path was anyways coming out to be less than 30, then, this *hold* requirement (of 30) will cause additional delay elements to be inserted, which will unnecessarily take up silicon area as well as power. So, designers need to align back the *hold* check edges. That means, the *hold* launch has to be moved forward by 3 cycles expressed in terms of the path start clock. This is achieved by completing the specification as:

set_multicycle_path –setup 4 –from clk1 –to clk2 -start
set_multicycle_path –hold 3 –from clk1 –to clk2 -start

This declaration is effectively telling that the *hold* launch edge is moved forward by 3 cycles (in terms of start clock). Figure 7.16 explains the movement of *hold* edges.

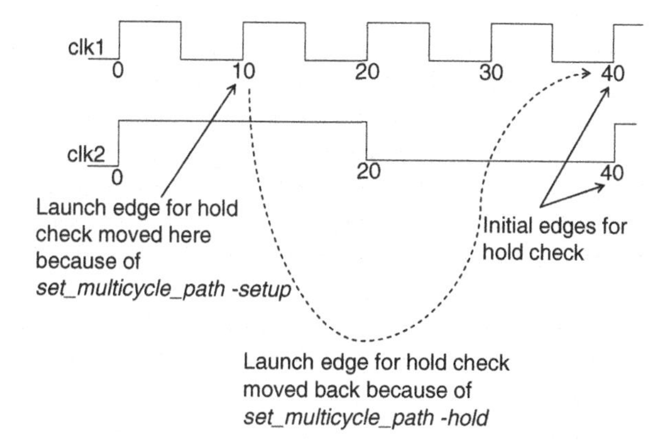

Fig. 7.16 Movement of launch edge for hold check

Now, you have achieved what you had wanted to convey, viz: the signal can have a delay of upto 40.

7.3.2.3 Protection Against Data Loss

This entire scheme is based on the fact that the data would be launched only once (the first cycle) in 4 cycles of start clock. Hence, *F1* needs to have control circuitry – which ensures that it launches the data only in the first cycle and then stops transmitting the data for 3 cycles, before launching the next data.

7.3.3 Protocol Based Data Transfer

Consider data being transmitted, where the sender and the destination clocks do not have an integral relationship. In such a case, the data transfer mechanism might depend on some protocol. Consider a simple protocol. The transmitter sends a signal *Ready*. This signal goes through 2 flops to be synchronized in the receiver domain, and then, it goes to an FSM. This FSM enables the receiver to capture the data. Figure 7.17 shows this simplified scheme.

After the data has been captured, there will be communication (*acknowledgement*) back to the transmitter. However, for this part of the discussion, we are

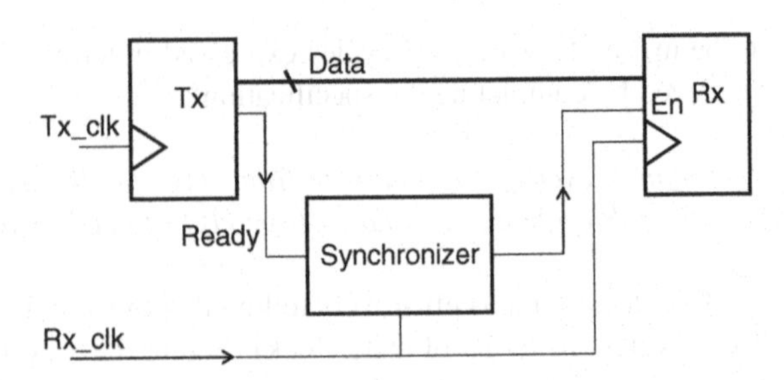

Fig. 7.17 Simple protocol for data transmission

not worried about the *acknowledgement*. Here, it is known that the signal *Ready* is going to take at least 2 cycles of destination clock, before it can enable the receiver. So, there is no need for the data itself to reach the receiver –within a single cycle. This is another example, where, you might want to use a *set_multicycle_path –setup*. And, since, the *–setup* will move the *hold* edge also, hence, a corresponding *set_multicycle_path –hold* should also be used, so that the *hold* edge comes back to its original location. The example commands would be

set_multicycle_path –setup 2 –from Tx_clk –to Rx_clk
set_multicycle_path –hold 1 –from Tx_clk –to Rx_clk

7.3.4 Multicycle Paths for False Paths

In the section on false paths, you saw that there were some paths, which need not be timed, even though the transitions can actually propagate along those paths. Some examples falling in this category include:

- Asynchronous clock domain crossings
- Quasi static signals
- Asynchronous control signals

Many designers prefer to specify a *set_multicycle_path*, instead of a *set_false_path*. Use of *set_multicycle_path* ensures an upper-bound, unlike *set_false_path* which simply does not have any limit at all.

7.3.5 Multicycle Paths for Deep Logic

So far, you saw various situations, where, you may declare a path as multi cycle, because for some reason, there is anyways no urgency for the data to be sampled – in the immediate next cycle itself. So, the command provides the freedom to the path to take its own time. However, sometimes, the data path has a deep cone of logic. For example, there might be a large arithmetic operator in the data path. Because of the delay through the path, the data cannot reach the destination within one cycle (resulting in violation of setup requirements). One option is obviously to reduce the clock frequency. But, that would mean slowing down the operation of the whole design, which is not always very desirable.

So, in such situations also, many times, you might declare such path as multi cycle path. Declaring the path as multi cycle will allow the data more than one cycle, so that the data can reach the destination. However, just declaring the path as multi cycle is not sufficient. The RTL design has to ensure that there are enough control mechanisms in place, so that:

- The destination does not capture any glitch
- The source does not transmit any new data till the previously transmitted data has been latched appropriately, else, the data might get lost or corrupted

The techniques described in earlier sections in this chapter can be used to ensure the above two requirements. Two of the most commonly used techniques for such situations include: (a) Handshake (or, a similar) mechanism to ensure a reliable data-transfer; or (b) Launch and Capture at specific cycles – as described in Section 7.3.1.3. By employing either of the mechanisms, you are effectively reducing the rate of data-transmittal across this specific path. (Alternately, you might introduce additional register elements in the path, in order to break the path into more segments, where, each segment is able to meet the requirements of a cycle. This is called *pipelining*. This does not involve any multicycle path, and, it does not reduce the frequency. However, not all situations allow pipelining.)

7.4 Conclusion

Timing exceptions should be used with great care. Some of the risks associated with incorrect usage/specification of timing exceptions include:

- A path getting under constrained. This effectively means, allowing more relaxation to the path, than it deserves. This will result in the risk of the path not meeting the timing. Thus, finally, the device will not be able to run at the desired frequency. This can happen due to any of the following situations.
  - A valid path getting declared as false-path.
  - A multi cycle path of N cycles declared as path with M cycles (where, $M > N$). A special case of this situation is, a single cycle path getting declared as multi-cycle.
- Multi cycle path declared with –setup, but, the hold edge not brought back to its original location. This can result in insertion of additional delay elements, which will consume both silicon real-estate and power.

- Glitch protection mechanism not put in place for multicycle path. This can result in functional failure of the device, if it captures glitches, or, transient values.
- A path getting over constrained. This effectively means a path which could have been allowed some relaxation, but, has not been allowed. This will result in higher effort of the implementation tools to meet the tighter timing. This can also potentially deteriorate the timing of the other paths, which are really of interest.

Hence, it is of utmost importance to apply exceptions carefully and with adequate consideration and validation. And, most importantly, as an RTL designer, you need to ensure that the RTL has adequate safeguards to prevent against data loss and glitch capture.

Chapter 8
Congestion

After the design is synthesized and scan insertion has been done, the cells are all placed. These cells are then connected. A lot of wires run all over the place in order to connect the respective pins of various cells with each other. Some common signals like clock, reset, power, ground also run all over the place in order to take the signals and supply or ground to all the cells.

At some places in the design, there are too many wires concentrated in a small area. This situation is called congestion. Once there is congestion, it becomes difficult to route more wire through this zone. As you can see, congestion is a purely back-end phenomenon, related to physical routing of interconnecting wires. Even though, congestion is a back-end phenomenon, there are certain characteristics of the RTL which could cause congestion. Hence, as an RTL designer, you should avoid these characteristics in your design – so that your design is free of congestion.

8.1 Impact of Congestion

Depending upon the severity of congestion, it might have different impact. In its most mild form, there might be congestion in a very localized zone. That means, all further interconnect wires should avoid this localized zone. Hence, some interconnects might have to be routed through a longer path, in order to avoid this congested zone. This in turn would mean higher capacitance on those wires as well as higher delay along those paths. Higher capacitance would mean higher drive strength for the driver – thus causing more power consumption. Also, higher delay along a path might not be affordable, if this path is a critical path.

In its most severe form, there might be too many locations within the design which are congested. In such a case, it might be difficult to realize the physical design, and, it might need some major changes – throwing the entire schedule for a toss. Sometimes, this might mean a larger die size – which could have impact on cost also, putting a question mark on the commercial viability itself. Before understanding the RTL characteristics that might impact congestion, it would help to get a brief introduction to the relevant portion of the physical design process.

S. Churiwala, S. Garg, *Principles of VLSI RTL Design*,
DOI 10.1007/978-1-4419-9296-3_8,

8.2 Physical Design Basics

The first step is Floor-plan. At this stage, all the blocks are assigned to some physical areas on the whole chip. The physical areas are assigned to the blocks, based on their sizes and the other blocks with which they are expected to interact. Blocks expected to interact more are kept closer together.

The next step is Cell-placement. In this stage, the individual cells are assigned specific locations within a block. Think of a block as consisting of large number of columns. All the cells would be placed within the column itself. These columns (and, cells within the columns) are all of fixed width. Any difference in cell-size is managed through the length along the column. Thus, a larger cell will require a longer space within the column. Even within the column, the cells are placed leaving spaces between them. For some methodologies, instead of columns, they use rows. Row or column is not important. The important thing is that the cells can all be placed only in a regular structure. Columns or rows are also referred as y-axis and x-axis respectively.

Once the cells are placed, all the routing wires, including power, ground, reset, clocks etc. are drawn to connect the respective pins. These wires can run only along tracks. So, if two wires are running along two adjoining tracks, no additional wire can be routed between these two wires. These wires run along x-axis or y-axis only, and, never diagonally. While drawing the interconnect wires, there is an obvious interest in keeping the routing distances to be minimal. All along these activities, timing is always kept in consideration. Sometimes, in order to meet timing, additional buffers might have to be inserted, or, certain gates might be replaced by their higher or lower drive versions (called, *upsizing* and *downsizing* respectively).

There are multiple layers of routing. Since any physical touch between wires means they get electrically connected, hence, in any given layer, all wires run parallel, and, in the next layer, they would run in the orthogonal direction. Wires on two different layers can be connected through *via*. So, anytime a wire has to change direction (from x-axis direction to y-axis direction or vice-versa), it has to go to a different layer, through a *via*. Vias have higher load, compared to normal interconnect wires. Hence, higher is the number of vias that a signal has to go through, it would need a bigger driver and will encounter more delays.

8.3 RTL Characteristics

As already mentioned, congestion is really a backend phenomenon. Once the gate level netlist is ready, it is possible to start getting reasonably good idea about the risk of congestion. However, even as an RTL designer you can play a role in reducing the risk of congestion – by avoiding certain characteristics in your design – which are likely to cause congestion.

8.3.1 High Utilization

Utilization refers to the ratio of physical area actually used by the gates of a block to the physical area reserved for the block. Even though, silicon real-estate is costly, still, it is not possible to utilize 100% of the silicon area to create devices on it. The gates are not all jam packed against each other. Rather, a good percentage of the silicon area is left open. This area is left open in order to allow for:

- routing of wires
- upsizing certain gates
- allowing several alternative locations for gates etc.

A higher utilization means a huge percentage of the available silicon area is already utilized for putting in the gates. This in turn means lesser space and lesser flexibility to route wires. This is very likely to create congestion. Figure 8.1a shows the middle column having closely packed gates – leaving space for just two tracks. Now, if a set of 3 wires need to be routed among two adjoining columns, they cannot be routed through this space, and, at least one wire has to be taken through a longer route.

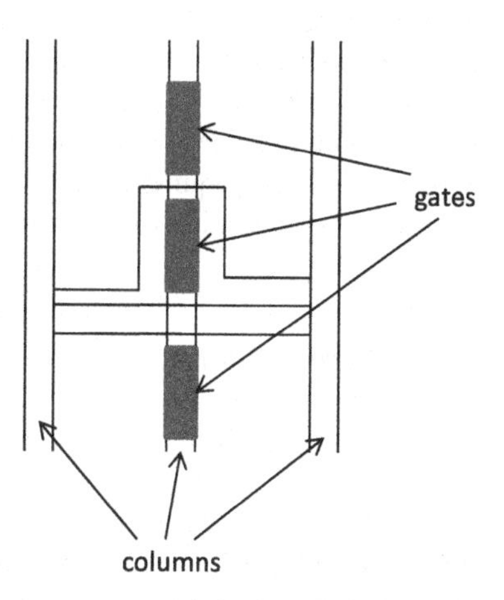

Fig. 8.1a Higher utilization: Wires need to be detoured

Figure 8.1b shows the middle column being less packed. Now, it can route the 3 wires directly, without the need for a detour.

So, obviously, a lower utilization has lesser tendency to create congestion. But, for a given physical area of gates, lower utilization means higher area for the block, i.e. wasted silicon real-estate. And, given that the real-estate on silicon is costly, you cannot expect a huge area reserved for your block – just so that you can have a low utilization. On the other hand, with higher utilization there is a risk of congestion. Congestion due to high utilization also means that the congestion is very likely to be

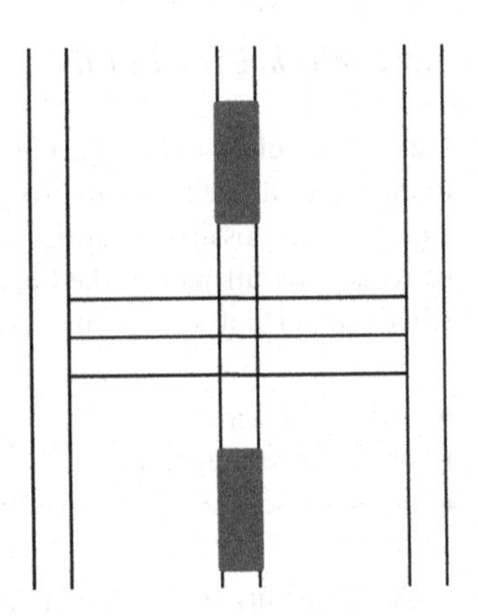

Fig. 8.1b Lower utilization: Detour avoided

wide-spread across the whole block, rather than being localized to specific locations. So, in practice, a good optimal utilization ratio is chosen.

Thus, as an RTL designer, you are expected to come up with a good estimate of the area that would be needed for your block, and, you should try to stick to that area. If the design that you finally realize takes higher area compared to your estimate, your block runs the risk of running into congestion. When you estimate your area, you should consider the timing and the power requirements also. More stringent timing means higher driver strength cells, which means bigger cells – and hence, higher area requirement. Besides the area required by the gates for the desired functionality, remember to add in the consideration for DFT requirements.

8.3.2 Large Macros

Consider a design, which has a huge mux – say, selecting between two busses each of 64 bits. Thus, 128 wires plus a select line need to enter the mux, and, another 64 wires need to come out of the mux's output. The location where the mux is placed is very likely to see a congestion – because 193 (128 data inputs, 1 select line and 64 data outputs) tracks in the immediate vicinity of the mux are occupied just by the signals of this mux. So, these tracks cannot be used by any other wire – which wants to connect two pins. Figure 8.2 shows this situation.

As an RTL designer, you should avoid using large arithmetic macros (adders, multipliers) or muxes, as wide data busses around these macros would cause congestion in the immediate vicinity. You should note that this congestion due to large macros is localized – in the immediate vicinity of the macro. Hence, even if the design level statistical data (say: utilization) indicate a low chance of congestion – there might still be a likelihood of local congestion in the immediate vicinity of the macro – even though, the rest of the design might be very clean – from congestion perspective.

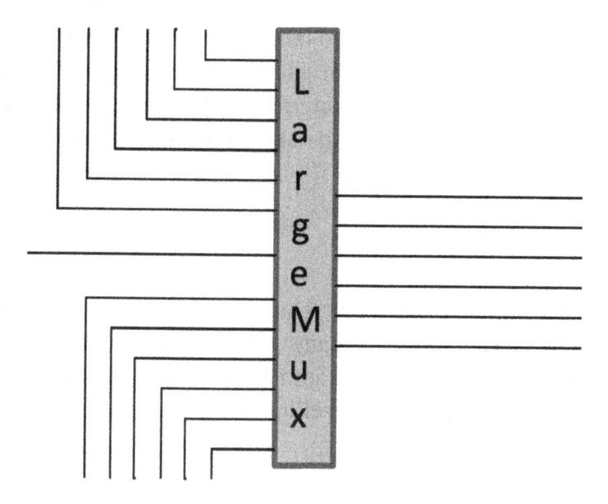

Fig. 8.2 Congestion around a large Mux

The phenomenon can also be explained using the concept of *Pin-density*. *Pin-density* refers to the number of pins per unit physical area. Since each pin will have a net connected to it, so, higher *pin density* means higher number of nets in a given area. Thus, higher *pin density* means higher chance of congestion in that specific location. Large macros effectively cause a very high local value of *pin-density*, even if the overall pin-density might not be high.

8.3.2.1 Composite Macro

Consider a large macro, having a complex functionality, with a large number of inputs. Similar to large data-path macros, such a composite macro with too many inputs will mean many signals coming in to feed into the large number of inputs. Thus, these many signals coming in to feed into the large number of inputs for the large composite macro could potentially cause congestion. Hence, during synthesis, you should avoid making use of very complex composite macro with a hugely complex boolean function with a large number of inputs. This appears slightly counter-intuitive. Smaller number of larger cells should mean lesser wires running all-around. So, that should be easier situation for routing. While, this is correct, the larger macros create local congestion.

8.3.3 Interaction with Many Blocks

Consider a portion of the design which interacts with too many components. Not all components can lie in close vicinity to this specific portion. This means, that this specific portion will have several wires which will travel long distances. Many wires travelling long distances means huge segments of many tracks occupied – leaving lesser rooms for other wires. Further, wires travelling long distances have longer delays and hence have to take the shortest possible path. This reduces the flexibility in routing them through alternate longer paths – if needed. So, what you have in effect is, many wires that are going to travel long distances and they have minimal flexibility in terms of alternative routes that they can take. Almost a sure shot recipe for inducing congestion!!

8.3.3.1 Wide Fanout

Extending the concept explained above, if a pin has a high fanout, it means it would have wires going to many other cells/pins. Since the fanout is large, not all the receivers will lie in the immediate vicinity. So, again, this is a situation of a huge number of wires – travelling long distances. Thus, large fanout can also induce congestion.

8.3.3.2 Wide Fanin Cone

A specific gate is fed by several inputs or registers. These registers and inputs feeding into the specific gate is referred as *fanin cone*. All the registers and inputs in the *fanin cone* of the specific gate will need to channelize their values through various gates, and, into this specific gate. Having a wide *fanin cone* means too many signals are trying to converge and merge on to the gate. Too many signals trying to merge together have a tendency to cause congestion. Thus, while writing your RTL, you should try to avoid having a very high *fanin or fanout cone*.

8.3.4 Too Many Critical Paths

The critical paths in a design have to be routed through the shortest possible route. This means, there is much lesser flexibility in routing the critical paths. If needed, they cannot be routed through an alternative less congested path. If your design has too many critical paths, then there are many paths which don't have flexibility. Though, these by themselves do not cause congestion problems, but, they can act as an impediment to alleviation of congestion – if it exists.

Figure 8.3a shows a typical timing profile. There are a small number of paths which are critical. For the rest of the paths, there is enough slack. These paths with high slack can be routed through alternative routes, and thus avoid areas of congestion.

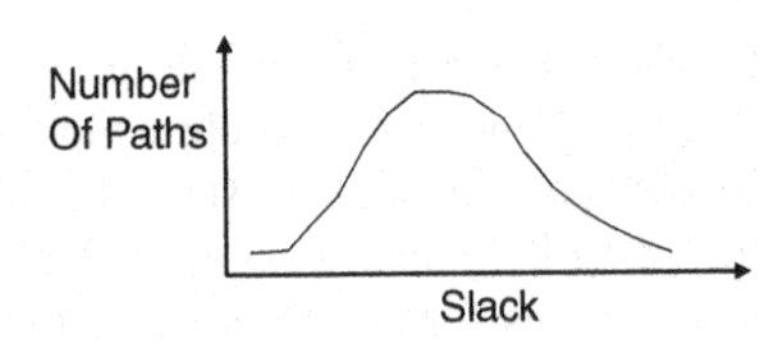

Fig. 8.3a Typical slack profile

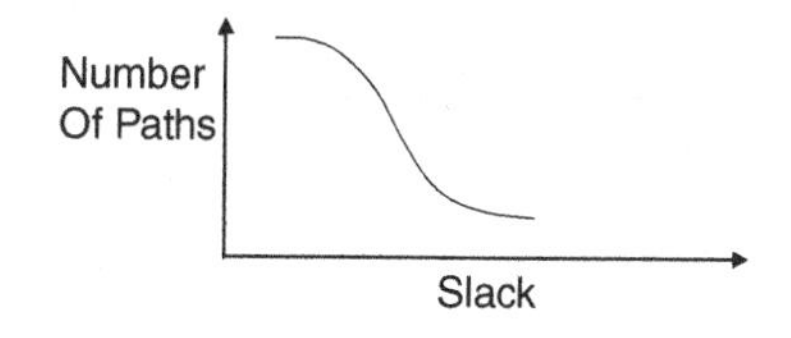

Fig. 8.3b Too many critical paths. Likely to cause congestion

Figure 8.3b shows a timing profile, where, too many paths are critical. This design is more likely to suffer congestion problems, as the routing tools cannot explore alternative routes.

8.4 Feedthrough

Sometimes, you have a *hard macro. Hard macro* refers to a portion of a design, which is already layout-complete. Hence, it cannot change its size, shape or any rearrangement of its internals. Since the hard macro acts as a blockage, hence, all the signals have to take a longer detour around it. So, if a hard macro sits between two interacting components, the signals between these two components have to go around the hard-macro. Figure 8.4 shows the resulting routing. This results in the possibility of congestion around the hard macro. Also, timing critical signals might not be able to afford the additional delay due to this detour.

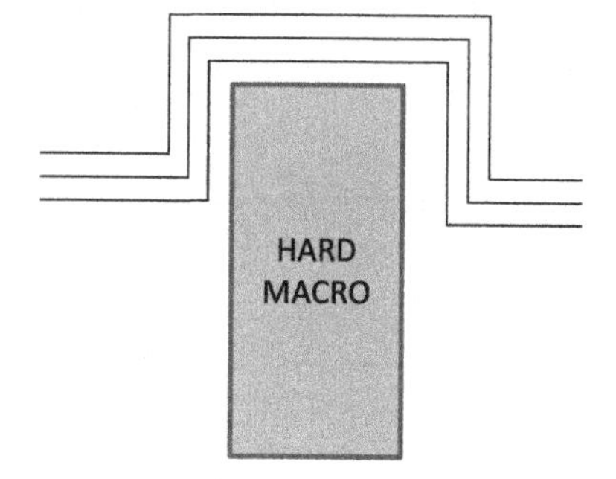

Fig. 8.4 Signals routed around hard macro

Thus, if you are developing a hard macro, you better keep the provision for a few *feedthroughs*. Feedthrough refers to a situation where an input of the hard macro connects to an output directly through a wire. So, any signal coming to this input will directly reach the output, without any processing within the hard macro. If the hard macro shown in Fig. 8.4 had some feedthroughs, the resulting routing would look as shown in Fig. 8.5

Note that the term *feedthrough* in this context has a different meaning compared to the meaning of the same term in Section 2.3. Though, it's the same term, its meaning is inferred based upon the context.

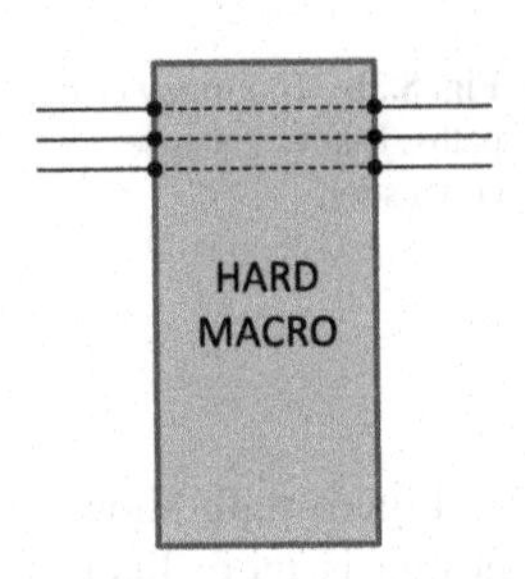

Fig. 8.5 Signals going through hard macro using feedthrough

8.5 Conclusion

Thus, even though, routing congestion is really a back-end phenomenon, some specific precautions at your end as an RTL designer could make it that much easier for your design to be taken through the complete process – including the back end steps for which you are not even responsible.

Appendix A
Interleaving of Processes

Section 5.4.2 of *IEEE Standard 1364–1995* (popularly called as *Verilog-95*) says that a simulator has a choice to suspend execution (even in the absence of any specific or explicit reason) of a partially completed block and move onto another concurrent event and come back to the current block later. As per the above interpretation, the following code excerpt is a race:

```
always @ (b or c)
a = b & c;

always @ (a or b or d)
if (a) o = b ^ d;
else o = b & d;
```

Consider the following three possible scenarios:

1. An event happens on *b*. The second ***always*** block gets triggered. The ***if*** statement is executed. It uses the old value of *a*. Simulator suspends the execution of this block, and, moves onto the first ***always*** block. New value of *a* is evaluated. But, since the second ***always*** block is already mid-way, hence, a change in *a* does not re-trigger this block (note the *blocking* assignment – which prevents the block from retriggering, till its execution is completed). So, the value of *o* is determined based on the old value of *a*.
2. An event happens on *b*. The second ***always*** block gets triggered. The ***if*** statement is executed, and the value of *o* is determined accordingly. Now, after this ***always*** block is completed, it comes to the first ***always*** block. The value of *a* is updated. Because of this event on *a*, the second ***always*** block gets triggered again. This time, the ***if*** condition has new value of *a*, and hence, the value of *o* is determined again based on the new value of *a*.
3. An event happens on *b*. The first ***always*** block gets triggered. After its completion, the second ***always*** block gets triggered. It sees the new value of *a*, and, the value of *o* is determined accordingly.

The only difference between scenarios 2 and 3 is the number of times the second ***always*** block gets triggered, but, the final values are still the same. However, in case

S. Churiwala, S. Garg, *Principles of VLSI RTL Design*,
DOI 10.1007/978-1-4419-9296-3, © Springer Science+Business Media, LLC 2011

of scenario 1, the actual value is also different. So, this is a race – because processes are allowed to get interleaved. Fortunately, even though, as per the LRM, the above is an example of a race; in practical world, these kinds of code segments are not considered as races; mainly because, so far, none of the simulators are known to perform this kind of interleaving unnecessarily. So, a simulator will execute either scenario 2 or scenario 3. In both cases, the results would be the same.

Thus, for all practical purposes, the above code segment is not a race, though, as per LRM – it may be classified as a race. In fact, many designers don't consider the above situation as a race.

Index